Materials for Catalysis in Energy

MATERIALS RESEARCH SOCIETY
SYMPOSIUM PROCEEDINGS VOLUME 1446

Materials for Catalysis in Energy

Symposium held April 9–13, 2012, San Francisco, California, U.S.A.

EDITORS

De-en Jiang
Oak Ridge National Laboratory
Oak Ridge, Tennessee, U.S.A.

Harold H. Kung
Northwestern University
Evanston, Illinois, U.S.A.

Rongchao Jin
Carnegie Mellon University
Pittsburgh, Pennsylvania, U.S.A.

Robert M. Rioux
The Pennsylvania State University
University Park, Pennsylvania, U.S.A.

Materials Research Society
Warrendale, Pennsylvania

Shaftesbury Road, Cambridge CB2 8EA, United Kingdom

One Liberty Plaza, 20th Floor, New York, NY 10006, USA

477 Williamstown Road, Port Melbourne, VIC 3207, Australia

314–321, 3rd Floor, Plot 3, Splendor Forum, Jasola District Centre, New Delhi – 110025, India

103 Penang Road, #05–06/07, Visioncrest Commercial, Singapore 238467

Cambridge University Press is part of Cambridge University Press & Assessment, a department of the University of Cambridge.

We share the University's mission to contribute to society through the pursuit of education, learning and research at the highest international levels of excellence.

www.cambridge.org
Information on this title: www.cambridge.org/9781605114231

Materials Research Society
506 Keystone Drive, Warrendale, PA 15086, USA
http://www.mrs.org

First published 2012

CODEN: MRSPDH

A catalogue record for this publication is available from the British Library

ISBN 978-1-605-11423-1 Hardback

CONTENTS

Preface vii

Materials Research Society Symposium Proceedings. ix

Hydrogen Generation for 500 hours by Photoelectrolysis of Water using GaN 1
W. Ohara, D. Uchida, T. Hayashi, M. Deura, and K. Ohkawa

Spray-deposited Co-Pi Catalyzed $BiVO_4$: A Low-cost Route Towards Highly Efficient Photoanodes. 7
Fatwa F. Abdi, Nienke Firet, Ali Dabirian, and Roel van de Krol

Artificial Photosynthesis - Use of a Ferroelectric Photocatalyst. 13
Steve Dunn and Matt Stock

Copper Tungstate ($CuWO_4$)–Based Materials for Photoelectrochemical Hydrogen Production. 19
Nicolas Gaillard, Yuancheng Chang, Artur Braun, and Alexander DeAngelis

Oxygen Reduction Reaction Electrocatalytic Activity of SAD-Pt/GLAD-Cr Nanorods 25
Wisam J. Khudhayer, Nancy Kariuki, Deborah J. Myers, Ali U. Shaikh, and Tansel Karabacak

Study of Co-assembled Conducting Polymers for Enhanced Ethanol Electro-oxidation Reaction. 33
Le Q. Hoa, Hiroyuki Yoshikawa, Masato Saito, and Eiichi Tamiya

Preparation and Characterization of Platinum/Ceria-based Catalysts for Methanol Electro-oxidation in Alkaline Medium 39
Christian L. Menéndez, Ana-Rita Mayol, and Carlos R. Cabrera

In Situ Spectroscopic Characterization of Some $LaNi_{1-x}Co_xO_3$ Perovskite Catalysts Active for CH_4 Reforming Reactions 47
Rosa Pereñiguez, Victor M. Gonzalez-Delacruz, Fatima Ternero, Juan P. Holgado, and Alfonso Caballero

Methane Combustion using CeO_2-CuO Fibers Catalysts 53
Felipe A. Berutti, Raquel P. Reolon,
Annelise K. Alves, and Carlos P. Bergmann

Role of Surface Oxide Layer during CO_2 Reduction at Copper Electrodes 59
Cheng-Chun Tsai, Joel Bugayong,
and Gregory L. Griffin

* **Porous Metal Oxides as Catalysts** 65
Boxun Hu, Christopher Brooks, Eric Kreidler,
and Steven L. Suib

Role of Pt Nanoparticles in Photoreactions on TiO_2 Photoelectrodes 71
Woo-Jin An, Wei-Ning Wang,
Balavinayagam Ramalingam, Somik Mukherjee,
Dariusz M. Niedzwiedzki, Shubhra Gangopadhyay,
and Pratim Biswas

Metal Oxides as Catalyst Promoters for Methanol Oxidation 77
Praveen Kolla, Kimberly Kerce, Hao Fong,
and Alevtina Smirnova

Supported Ni Catalyst Made by Electroless Ni-B Plating for Diesel Autothermal Reforming 83
Zetao Xia, Liang Hong, Wei Wang,
and Zhaolin Liu

Shape-controlled Synthesis of Silver and Palladium Nanocrystals using β-Cyclodextrin 89
Gilles Berhault, Hafedh Kochkar,
and Abdelhamid Ghorbel

Author Index 95

Subject Index 97

*Invited Paper

PREFACE

Growing demand for energy and the need to reduce our society's carbon footprint calls for transformative measures to increase efficiency in energy consumption and sustainable methods of energy production and storage. Novel materials will be key to these transformative technologies by acting as catalysts and facilitating desired chemical transformations. It is becoming increasingly evident that the integration of materials science perspectives into catalyst discovery, synthesis, and characterization will provide new opportunities for novel catalytic materials in energy-related applications.

Symposium U of the 2012 MRS Spring Meeting in San Francisco, California, "Materials for Catalysis in Energy," was held April 10–13. The objective of the organizers was to bring together researchers from materials science, chemical synthesis, heterogeneous catalysis, electrocatalysis, and photocatalysis to highlight recent progress and discuss challenges and opportunities in the materials aspect of catalysis research and development for energy applications. This symposium was the second one in the recent history of MRS meetings with a particularly strong focus on the materials aspect of catalysis; the first one was pioneered and organized by Sheng Dai, Harold H. Kung, Jun Liu, and Chung-Yuan Mou at the 2009 MRS Fall Meeting in Boston, Massachusetts.

Close to 200 abstracts were received and about 150 papers were presented in Symposium U in Spring 2012, demonstrating the significant interest from the broader catalysis community in this topical area. This volume contains a cross section of papers presented in the symposium, and highlights the interdisciplinary nature of this research area. We thank the authors, the reviewers, and the MRS staff in making this volume possible and we hope you enjoy reading these outstanding contributions.

De-en Jiang
Harold H. Kung
Rongchao Jin
Robert M. Rioux

July 2012

MATERIALS RESEARCH SOCIETY SYMPOSIUM PROCEEDINGS

Volume 1426 — Amorphous and Polycrystalline Thin-Film Silicon Science and Technology – 2012, B. Yan, H. Gleskova, C.C. Tsai, T. Sameshima, J.D. Cohen, ISBN 978-1-60511-403-3
Volume 1427E — Heterogeneous Integration Challenges of MEMS, Sensor and CMOS LSI, 2012, K. Masu, K. Sawada, H. Toshiyoshi, B. Charlot, A.P. Pisano, ISBN 978-1-60511-404-0
Volume 1428E — Interconnect Challenges for CMOS Technology–Materials, Processes and Reliability for Downscaling, Packaging and 3D Stacking, 2012, G. Dubois, F. Iacopi, A. Sekiguchi, S.W. King, C. Dussarrat, ISBN 978-1-60511-405-7
Volume 1429E — Nanocontacts–Emerging Materials and Processing for Ohmicity and Rectification, 2012, A.A. Talin, M.S. Islam, C. Lavoie, K-N. Tu, ISBN 978-1-60511-406-4
Volume 1430 — Materials and Physics of Emerging Nonvolatile Memories, 2012, Y. Fujisaki, P. Dimitrakis, E. Tokumitsu, M.N. Kozicki, ISBN 978-1-60511-407-1
Volume 1431E — Phase-Change Materials for Memory and Reconfigurable Electronics Applications, 2012, P. Fons, B.J. Kooi, B-S. Lee, M. Salinga, R. Zhao, ISBN 978-1-60511-408-8
Volume 1432 — Reliability and Materials Issues of III-V and II-VI Semiconductor Optical and Electron Devices and Materials II, 2012, O. Ueda, M. Fukuda, K. Shiojima, E. Piner, ISBN 978-1-60511-409-5
Volume 1433E — Silicon Carbide 2012 – Materials, Processing and Devices, 2012, F. Zhao, E. Sanchez, H. Tsuchida, R. Rupp, S.E. Saddow, ISBN 978-1-60511-410-1
Volume 1434E — Recent Advances in Superconductors, Novel Compounds and High-Tc Materials, 2012, J. Shimoyama, E. Hellstrom, M. Putti, K. Matsumoto, T. Kiss, ISBN 978-1-60511-411-8
Volume 1435E — Organic and Hybrid-Organic Electronics, 2012, P. Blom, O. Hayden, J. Park, H. Richter, F. So, ISBN 978-1-60511-412-5
Volume 1436E — Advanced Materials and Processes for Systems-on-Plastic, 2012, T. Someya, I. McCulloch, T. Takenobu, I. Osaka, S. Steudel, A.C. Arias, ISBN 978-1-60511-413-2
Volume 1437E — Group IV Photonics for Sensing and Imaging, 2012, K. Ohashi, R.A. Soref, G. Roelkens, H. Minamide, Y. Ishikawa, ISBN 978-1-60511-414-9
Volume 1438E — Optical Interconnects–Materials, Performance and Applications, 2012, E. Suhir, D. Read, R. Houbertz, A.M. Earman, ISBN 978-1-60511-415-6
Volume 1439 — Nanowires and Nanotubes–Synthesis, Properties, Devices, and Energy Applications of One-Dimensional Materials, 2012, J. Motohisa, L.J. Lauhon, D. Wang, T. Yanagida, ISBN 978-1-60511-416-3
Volume 1440E — Next-Generation Energy Storage Materials and Systems, 2012, D. Qu, ISBN 978-1-60511-417-0
Volume 1441E — Advanced Materials and Nanoframeworks for Hydrogen Storage and Carbon Capture, 2012, M. Fichtner, ISBN 978-1-60511-418-7
Volume 1442E — Titanium Dioxide Nanomaterials – 2012, 2012, X. Chen, G. Tulloch, C. Li, ISBN 978-1-60511-419-4
Volume 1443E — Bandgap Engineering and Interfaces of Metal Oxides for Energy, 2012, J.D. Baniecki, S. Zhang, G. Eres, N. Valanoor, W. Zhu, ISBN 978-1-60511-420-0
Volume 1444 — Actinides and Nuclear Energy Materials, 2012, A.D. Andersson, C.H. Booth, P.C. Burns, R. Caciuffo, R. Devanathan, T. Durakiewicz, M. Stan, V. Tikare, S.W. Yu, ISBN 978-1-60511-421-7
Volume 1445E — Bioinspired Materials for Energy Applications, 2012, B. Schwenzer, E.D. Haberer, B-L. Su, Y.J. Lee, D. Zhang, ISBN 978-1-60511-422-4
Volume 1446 — Materials for Catalysis in Energy, 2012, D. Jiang, R. Jin, R.M. Rioux, ISBN 978-1-60511-423-1
Volume 1447 — Nanostructured and Advanced Materials for Solar-Cell Manufacturing, 2012, "B. Nelson, L. Tsakalakos, A. Salleo, S. Mukhopadhyay, U. Bach, L. Schmidt-Mende, T. Brown, A. Fontcuberta i Morral, M. Law, ISBN 978-1-60511-424-8
Volume 1448E — Conjugated Organic Materials for Energy Conversion, Energy Storage and Charge Transport, 2012, L.P. Yu, ISBN 978-1-60511-425-5
Volume 1449 — Solution Synthesis of Inorganic Films and Nanostructured Materials, 2012, M. Jain, X. Obradors, Q.X. Jia, R.W. Schwartz, ISBN 978-1-60511-426-2
Volume 1450E — Hierarchically Self-Assembled Materials–From Molecule to Nano and Beyond, 2012, C. Li, ISBN 978-1-60511-427-9
Volume 1451 — Nanocarbon Materials and Devices, 2012, J. Appenzeller, M.J. Buehler, Y. Homma, E.I. Kauppinen, K. Matsumoto, C.S. Ozkan, N. Pugno, K. Wang, ISBN 978-1-60511-428-6

Materials Research Society Symposium Proceedings

Volume 1452E — Nanodiamond Particles and Related Materials–From Basic Science to Applications, 2012, O. Shenderova, E. Osawa, A. Krueger, ISBN 978-1-60511-429-3
Volume 1453 — Functional Hybrid Nanoparticles and Capsules with Engineered Structures and Properties, 2012, N.S. Zacharia, J.B. Tracy, Y. Yin, U. Jeong, D.H. Kim, C.J. Martinez, Z. Lin, K. Benkstein, ISBN 978-1-60511-430-9
Volume 1454 — Nanocomposites, Nanostructures and Heterostructures of Correlated Oxide Systems, 2012, T. Endo, N. Iwata, H. Nishikawa, A. Bhattacharya, L.W. Martin, ISBN 978-1-60511-431-6
Volume 1455E — Nanoscale Materials Modification by Photon, Ion, and Electron Beams, 2012, Y. Shinozuka, T. Kanayama, R.F. Haglund, Jr., F. Träger, ISBN 978-1-60511-432-3
Volume 1456E — Nanoscale Thermoelectrics 2012 – Materials and Transport Phenomena, 2012, R. Venkatasubramanian, A. Boukai, T. Borca-Tasciuc, K. Koumoto, L. Chen, ISBN 978-1-60511-433-0
Volume 1457E — Plasmonic Materials and Metamaterials, 2012, L.A. Sweatlock, J.A. Dionne, V. Kovanis, J. van de Lagemaat, ISBN 978-1-60511-434-7
Volume 1458E — New Trends and Developments in Nanomagnetism, 2012, H.P. Oepen, A. Berger, P. Fischer, K. Koike, ISBN 978-1-60511-435-4
Volume 1459E — Topological Insulators, 2012, H. Buhmann, X.-L. Qi, P. Hofmann, ISBN 978-1-60511-436-1
Volume 1460E — DNA Nanotechnology, 2012, M. Mertig, H. Yan, I. Willner, H. Dietz, ISBN 978-1-60511-437-8
Volume 1461E — Structure-Function Design Strategies for Bio-Enabled Materials Systems, 2012, V.T. Milam, H. Bermudez, R.R. Naik, M. Knecht, ISBN 978-1-60511-438-5
Volume 1462E — Manipulating Cellular Microenvironments, 2012, R. Bashir, ISBN 978-1-60511-439-2
Volume 1463E — Mechanobiology of Cells and Materials, 2012, M. Butte, D. Gourdon, M. Smith, S. Diez, ISBN 978-1-60511-440-8
Volume 1464E — Molecules to Materials–Multiscale Interfacial Phenomena in Biological and Bio-Inspired Materials, 2012, E. Spoerke, D. Joester, E. Sone, R. Bitton, T. Kelly, ISBN 978-1-60511-441-5
Volume 1465E — Structure/Property Relationships in Biological and Biomimetic Materials at the Micro-, Nano- and Atomic-Length Scales, 2012, P. Zaslansky, B. Pokroy, N. Tamura, S. Habelitz, ISBN 978-1-60511-442-2
Volume 1466E — Interfaces in Materials, Biology and Physiology, 2012, A.M.K. Esawi, G.M. Genin, T.J. Lu, U.G.K. Wegst, J.J. Wilker, ISBN 978-1-60511-443-9
Volume 1467E — Integration of Natural and Synthetic Biomaterials with Organic Electronics, 2012, M. Irimia-Vladu, C.J. Bettinger, L. Torsi, J. Rogers, ISBN 978-1-60511-444-6
Volume 1468E — Nanomedicine for Molecular Imaging and Therapy, 2012, X. Chen, J. Cheon, M. Amiji, S. Nie, ISBN 978-1-60511-445-3
Volume 1469E — Plasma Processing and Diagnostics for Life Sciences, 2012, M. Hori, A. Fridman, P. Favia, N. Itabashi, M. Shiratani, ISBN 978-1-60511-446-0
Volume 1470E — Computational Materials Design in Heterogeneous Systems, 2012, M. Sushko, D. Quigley, O. Hod, D. Duffy, ISBN 978-1-60511-447-7
Volume 1471E — Rare-Earth-Based Materials, 2012, J. Dickerson, ISBN 978-1-60511-448-4
Volume 1472E — Transforming Education in Materials Science and Engineering, 2012, A-R. Mayol, M.M. Patterson, ISBN 978-1-60511-449-1
Volume 1473E — Functional Materials and Ionic Liquids, 2012, S. Dai, T.P. Lodge, R.D. Rogers, P. Wasserscheid, M. Watanabe, ISBN 978-1-60511-450-7
Volume 1474E — Local Probing Techniques and In-Situ Measurements in Materials Science, 2012, N. Balke, H. Wang, J. Rijssenbeek, T. Glatzel, ISBN 978-1-60511-451-4

Prior Materials Research Society Symposium Proceedings available by contacting Materials Research Society

Mater. Res. Soc. Symp. Proc. Vol. 1446 © 2012 Materials Research Society
DOI: 10.1557/opl.2012.1249

Hydrogen Generation for 500 hours by Photoelectrolysis of Water using GaN

W. Ohara, D. Uchida, T. Hayashi, M. Deura, and K. Ohkawa
Department of Applied Physics, Tokyo University of Science, 1-3 Kagurazaka, Shinjuku, Tokyo 162-8601, Japan

ABSTRACT

We confirmed that GaN photocatalyst with NiO cocatalyst (GaN-NiO) continuously produced hydrogen from water for 500 hours without any extra bias. The GaN-NiO photocatalyst was hardly etched and 184-mL hydrogen gas was produced from the electric charge of 1612 coulombs, the Faradic efficiency of which was 89.2%. The conversion efficiency from incident light energy to hydrogen chemical energy was 0.98% in average for 500 h. The incident photon-to-current conversion efficiency (IPCE) was 50% at 300 nm and 35% at 350 nm after the experiment, which was much higher than those of other semiconductor-based photocatalysts.

INTRODUCTION

Since no CO_2 is discharged after its combustion, hydrogen is promising as the new energy source instead of fossil fuels. As a method of H_2 generation, photoelectrolysis of water using photocatalysts is attracting because only water and solar light energy are necessary [1-5].

We have focused on III-nitrides, especially GaN, as the photocatalyst [6-14]. Their band edge potentials are comparatively higher than those of conventional oxides used as photocatalysts and straddle the oxidation-reduction level of water. In fact, we have realized H_2 generation using GaN photocatalyst without any extra bias. Moreover, they can absorb not only ultraviolet but also visible light using InGaN alloys by changing the group-III content.

However, most of the photocatalysts have insufficient durability for practical use. Although GaN is chemically stable essentially, GaN layer itself is etched during photocatalytic reaction due to holes generated by light illumination and accumulated on the surface of the GaN layer. This etching reduces the durability of the photocatalyst. We have found that deposition of NiO cocatalyst on the GaN layer avoids the etching [15]. Furthermore, the amount of H_2 generated for GaN photocatalyst with NiO (GaN-NiO) increased more than four times compared to that for GaN (GaN w/o NiO). In this study, we investigated the stability of hydrogen generation from water using GaN-NiO for 500 hours.

EXPERIMENT

We used a 3-µm-thick n-type GaN layer on a sapphire substrate grown by metalorganic vapor-phase epitaxy, and NiO was deposited with around 1.2% coverage on the GaN surface. The film thickness of GaN was 3.1 µm, and its room temperature carrier density and electron mobility were 1.2×10^{17} cm^{-3} and 580 $cm^2/V\cdot s$, respectively. This GaN-NiO working electrode connected to the Pt counterelectrode was dipped into a NaOH solution with the concentration of 1 mol/L. The irradiated area of the GaN-NiO electrode was adjusted to 1.0 cm^2 by covering the remaining area with epoxy resin. A Xe lamp with the energy density of 100 mW/cm^2 was used as the light source and the photocurrent density was measured using a potentiostat. All of the experiments were performed at room temperature, and no bias was applied to this system. We conducted 500 h experiment with 50 repetitions of a 10 h experiment. We measured the amount of generated H_2, the energy conversion efficiency, the incident photon-to-current conversion efficiency (IPCE), and change in the surface of the GaN layer.

DISCUSSION

Time evolution of the amount of hydrogen

Time evolution of the amount of generated hydrogen for 500 hours is shown in Fig. 1. The amount was increased gradually until 60h. This may be due to the oxidation of epoxy resin instead of O_2 generation [16]. Drastic drop between 110 and 120 h is caused by degradation of the intensity of the Xe lamp. In fact, the amount was recovered after the lamp was exchanged after 480h. Therefore, the decrease at120 h is not due to the deterioration of the GaN-NiO electrode. Since the amount of generated H_2 was stable between 120 to 480 h, the function as the photocatalyst retained for 500 h. Total amount of H_2 produced for 500 h was 184 mL from the electric charge of 1612 C, and thus the Faradic efficiency (electric charges used for hydrogen generation/ generated electric charges) was 89.2%. This amount of generated H_2 corresponds to 37 $mL/cm^2/100$ h, which is over one magnitude larger than that for Cu_2O (1.6×10^{-4} $mL/cm^2/100$ h) [17] or $NiO/NaTaO_3$:La (2.2 $mL/cm^2/100$ h) [18].

The surface morphology of the GaN-NiO and GaN w/o NiO electrode was observed by a Nomarski microscope as shown in Fig.2. A lot of dark points were observed over the entire surface of GaN w/o NiO after the experiment. All of these points were pits, which were measured by a stylus surface profiler. The value of RMS (root mean square) of the surface roughness was 101 nm, which was 100 times larger than that before the experiment. Ga, N, and

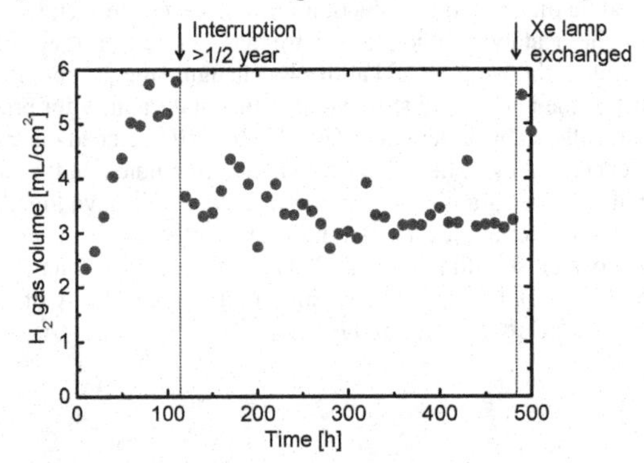

Figure 1. Time evolution of generated hydrogen for 500 hours.

O were detected in the points by energy dispersive X-ray spectrometry (EDS). Therefore, the dark points were due to the etching of the surface. In contrast, the GaN-NiO surface was not clearly changed even after 500 h, and the RMS of the roughness was 9.3 nm. Although a few dark points were observed on the surface as shown in Fig. 2(b), the amount of generated H_2 was constant; this is because ratio of the degraded area was small compared to the whole surface area. Some of these points were not pits but projections. Since the C peak was detected in addition to the Ga, N, and O peaks in the dark points of GaN-NiO, the projected points were maybe attributed to some organic decomposition from such as epoxy resin.

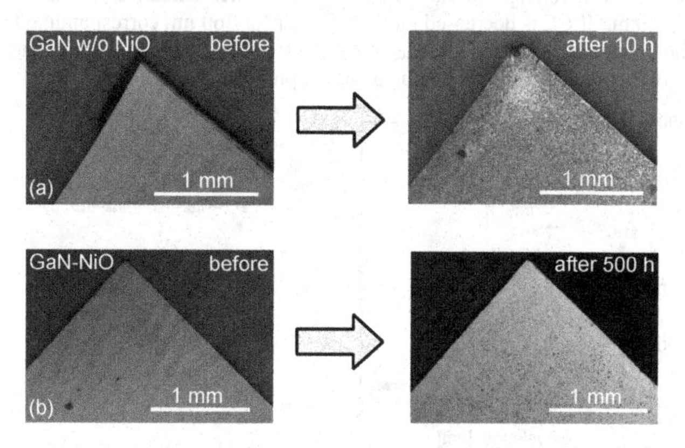

Figure 2. The surface morphology of the (a) GaN w/o NiO and (b) GaN-NiO electrode observed by a Nomarski microscope.

Energy conversion efficiency

The energy conversion efficiency from incident light energy to hydrogen chemical energy was described by Eq. 1.

$$\text{Energy convertion efficiency} = \frac{\text{Hydrogen energy}}{\text{Incidence light energy}} = \frac{\Delta G^0 \times n}{P \times A \times t} \qquad (1)$$

where ΔG^o=237 kJ/mol is the Gibbs energy of H_2 combustion, n is the number of hydrogen molecules, P is the incident light intensity, A is the area of light illumination, and t is time. In the case of the GaN-NiO photocatalyst, this efficiency was 1.54% in maximum (100-110 h) and 0.98% in average for 500 h. This efficiency was 4.6 times larger than that for GaN w/o NiO of 0.33%.

Figure 3 is the IPCE of the GaN-NiO photocatalyst before and after the 500 h experiment. We used a monochromatic light with full-width at half maximum of 2.8 nm at 350 nm from a Xe lamp using a monochromator, and incident light energy to the GaN surface was a few hundred μW. IPCE is the conversion efficiency from the number of photons to the number of electrons produced as defined in Eq. 2.

$$\mathrm{IPCE} = \frac{\text{number of electrons produced}}{\text{number of incident photons}} = \frac{I_P/e}{P \times \lambda/hc} \tag{2}$$

where I_P is the photocurrent density, e is the elementary electric charge, P is the power density of monochromatic light, λ is wavelength of the incident light, h is the Planck's constant, and c is the speed of light in vacuum. IPCE is decreased rapidly above near 360 nm corresponding to 3.42 eV (363 nm) of the bandgap of GaN. ICPE was 50% at 300 nm and 35% at 350 nm, and these values were much higher than other semiconductor-based photocatalysts.

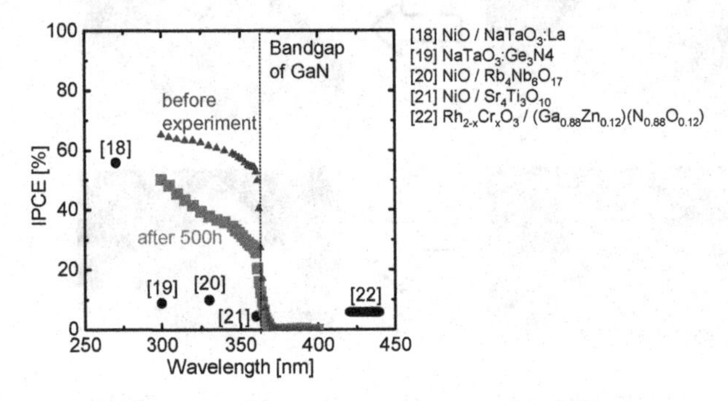

Figure 3. The IPCE of the GaN-NiO photocatalyst before and after the 500 h experiment.

CONCLUSIONS

We investigated the durability of hydrogen generation from water using the GaN-NiO photocatalyst. GaN-NiO produced H_2 continuously with high Faradic efficiency for 500 h without heavy etching. The amount of generated H_2 was over one magnitude larger than that using other materials. The energy conversion efficiency and the ICPE were both distinctly higher than those for other photocatalysts. In conclusion, GaN-NiO works as the photocatalyst for hydrogen generation from water with high efficiency and good stability.

REFERENCES

1. A. Fujishima and K. Honda, Nature **238**, 37 (1972).
2. D. Jing, L. Guo, L. Zhao, X. Zhang, H. Liu, M. Li, S. Shen, G. Liu, X. Hu, X. Zhang, K. Zhang, L. Ma, P. Guo, Int. J. Hydrogen Energy **35**, 7087 (2010).
3. B.Kraeutler and A. J. Bard, J. Am. Cham. Soc. **100**, 4317 (1978).
4. R. Memming, *Semiconductor Electrochemistry*, WILEY-VCH, p. 105 (2001).

5. S. S. Kocha, M. W. Peterson, D. J. Arent, J. M. Redwing, M. A. Tischler, J. A. Turner, J. Electrochem. Soc. **142**, L236 (1995).
6. M. Ono, K. Fujii, T. Ito, Y. Iwaki, T. Yao, K. Ohkawa, J. Chem. Phys. **126**, 054708 (2007).
7. K. Fujii and K. Ohkawa, Jpn. J. Appl. Phys. **44**, L909 (2005).
8. K. Fujii, K. Kusakabe, K. Ohkawa, Jpn. J. Appl. Phys. **44**, 7433 (2005).
9. K. Fujii and K. Ohkawa, J. Electrochem. Soc. **153**, A468 (2006).
10. K. Fujii and K. Ohkawa, Phys. Status Soldi C **3**, 2270 (2006).
11. K. Fujii, M. Ono, T. Ito, K. Ohkawa, Mater. Res. Soc. Sym. Proc. 0885-A11-04.1 (2006).
12. K. Fujii, T. Ito, M. Ono, Y. Iwaki, T. Yao, K. Ohkawa, Phys. Status Solidi C **4**,2650 (2007).
13. K. Fujii, Y. Iwaki, H. Masui, T. J. Baker. M. Iza, H. Sato, J. Keading, T. Yao, J. S. Speck, S. P. DenBaars, S. Nakamura, K. Ohkawa, Jpn. J. Appl. Phys. **46**, 6573 (2007).
14. K. Fujii, T. Karasawa, K. Ohkawa, Jpn. J. Appl. Phys. **44**, 543 (2005).
15. F. Sano, T. Koyama, M. Sorimachi, A. Hirako, K. Ohkawa, The 8th Int'l Conf. on Nitride Semiconductors, p. 18 (2009).
16. K. Ito, S. Ikeda, M. Yoshida, S. Ohta, and T. Iida, Bull. Chem. Soc. Jpn. **57**, 583 (1984).
17. M. Hara, T. Kondo, M. Komoda, S. Ikeda, K. Shinohara, A. Tanaka, J. N. Kondo, K. Domen, Chem. Commun. **10**, 1039 (1998).
18. H. Kato, K. Asakura, A. Kudo, J. Am. Chem. Soc. **125**, 3082 (2003).
19. H. Kato and A. Kudo, J. Phys. Chem. B **105**, 4285 (2001).
20. K. Sayama and H.Arakawa, Catal. Today **28**, 175 (1997).
21. Y. G. Ko and W. Y. Lee, Catal. Lett. **83**, 157 (2002).
22. K. Maeda, K. Teramura, K. Domen, J. Catal. **254**, 198 (2008).

Mater. Res. Soc. Symp. Proc. Vol. 1446 © 2012 Materials Research Society
DOI: 10.1557/opl.2012.811

Spray-deposited Co-Pi Catalyzed $BiVO_4$: a low-cost route towards highly efficient photoanodes

Fatwa F. Abdi, Nienke Firet, Ali Dabirian and Roel van de Krol
Materials for Energy Conversion and Storage (MECS), Department of Chemical Engineering, Delft University of Technology, P.O. Box 5045, 2600 GA Delft, The Netherlands

ABSTRACT

Bismuth vanadate ($BiVO_4$) thin films are deposited by a low-cost and scalable spray pyrolysis method. Its performance under AM1.5 illumination is mainly limited by slow water oxidation kinetics. We confirm that cobalt phosphate (Co-Pi) is an efficient water oxidation catalyst for $BiVO_4$. The optimum thickness of $BiVO_4$ is 300 nm, resulting in an AM1.5 photocurrent of 1.9 mA/cm^2 at 1.23 V vs. RHE when catalyzed with Co-Pi. Once the water oxidation limitation is removed, the performance is limited by low charge separation efficiency. This causes more than 60% of the electron-hole pairs to recombine before reaching the respective interfaces. The slow electron transport is shown to be the main cause of this low efficiency, and future efforts should therefore be focused on addressing this key limitation.

INTRODUCTION

$BiVO_4$ is considered to be a promising photoanode material for solar water splitting applications. The monoclinic phase has a bandgap of 2.4 eV [1], and can absorb up to 11% of the solar spectrum. Additionally, its conduction band edge is located very close to the reversible hydrogen electrode (RHE) level [2], which enables water splitting at modest external bias potentials. Recent efforts on this material have resulted in significant performance enhancements by the application of oxygen evolution catalysts [3-8]. However, little is known about the factor(s) that limit the performance of $BiVO_4$ after the slow water oxidation kinetics is addressed.

In this study, we use a low-cost spray pyrolysis technique to deposit high-quality $BiVO_4$ films. This technique is highly scalable, an important requirement for large-scale production of solar energy conversion devices. To illustrate this, full conversion to a solar-driven society in 2050, based on 10% efficient devices, requires production rates in the order of 500 m^2/s. This is just one order of magnitude more than the global spray deposition rate in the automotive industry (15 m^2/s). We show that 300 nm is the optimum thickness for our spray-deposited $BiVO_4$ films, resulting in an AM1.5 photocurrent of 1.9 mA/cm^2 at 1.23 V vs. RHE when catalyzed with cobalt phosphate (Co-Pi). This photoanode is limited by a poor charge separation efficiency (η_{sep} < 0.4), which is caused by the inherently slow electron transport in $BiVO_4$.

EXPERIMENTAL DETAILS

Dense thin films of $BiVO_4$ were prepared by spray pyrolysis, as illustrated in Fig. 1a. A $BiVO_4$ precursor solution was prepared by dissolving $Bi(NO_3)_3.5H_2O$ (98%, Alfa Aesar) in acetic acid (98%, Sigma Aldrich) and $VO(AcAc)_2$ (99%, Alfa Aesar) in absolute ethanol (Sigma Aldrich). The Bi solution was then added to the V solution, and the mixture was diluted to 4 mM with excess ethanol. The substrates are FTO-coated glass (15 Ω/□, TEC-15, Hartford Glass Co.), which had been cleaned by three successive 15 min. ultrasonic rinsing steps in 10 vol% Triton®, acetone and ethanol. The substrates were placed on a heating plate that was set to 450°C during deposition. The spray nozzle (Quickmist Air Atomizing Spray) was placed 20 cm above the heating plate, and was fed by two nitrogen gas lines (labeled 1 and 2 in Fig. 1) and the liquid $BiVO_4$ precursor solution. Figure 1b shows the schematic representation of the spray nozzle. A pulsed deposition mode was used, with one spray cycle consisting of 5 seconds of spray time and 55 seconds of delay time to allow the solvent to evaporate. In order to prevent large droplets from falling onto the substrate when switching the $BiVO_4$ flow on or off via line 1, gas was flown through line 2 before and after starting and stopping line 1 (Fig. 1c). The thicknesses of $BiVO_4$ were controlled by the number of spray cycles, with a deposition rate of ~1 nm per cycle. Prior to $BiVO_4$ deposition, ~80 nm of SnO_2 layer was deposited onto FTO substrate to prevent recombination of electrons and holes at the FTO/$BiVO_4$ interface [9,10]. After the deposition, the SnO_2/$BiVO_4$ samples were annealed for 2 hours at 450°C in air to further improve the crystallinity.

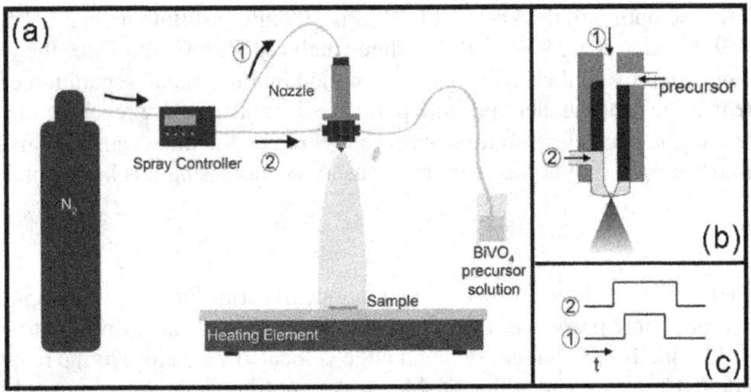

Figure 1. (a) A schematic diagram of the spray pyrolysis setup. The arrows labeled 1 and 2 represent two separate nitrogen flows towards the spray nozzle. (b) A schematic representation of the spray nozzle. (c) Timing of the two separate spray pulses that are supplied to lines 1 and 2.

A 30 nm Co-Pi catalyst was electrodeposited onto the surface of $BiVO_4$ in an electrochemical cell using a three-electrode configuration, according to the recipe from Nocera et al [11]. The electrolyte is made by dissolving 0.5 mM $Co(NO_3)_2$ (99%, Acros Organics) in a 0.1 M KPi solution (pH ~7). The potential of the working electrode was controlled by a potentiostat (EG&G PAR 283). A coiled Pt wire and an Ag/AgCl electrode (XR300, saturated KCl and AgCl solution, Radiometer Analytical) were used as the counter and reference electrodes, respectively. The electrodeposition was carried out at a constant voltage of 1.3 V_{NHE} (1.7 V_{RHE}) for 15 minutes.

Photoelectrochemical characterization was carried out using the same three-electrode configuration. The electrolyte used is an aqueous 0.5 M K_2SO_4 (99%, Alfa Aesar) solution buffered to pH ~5.6 with K_2HPO_4/KH_2PO_4. The light source is a Newport Sol3A Class AAA Solar Simulator (type 94023A-SR3) that provides simulated AM1.5 illumination (100 mW/cm^2).

Photochemical stability measurements were performed using monoclinic $BiVO_4$ powder (Ciba Specialty Chemicals), with a typical particle size of 200-500 nm. This powder was dispersed in aqueous solutions with a pH ranging from 1 - 13 (adjusted using HCl/KOH), and stirred for 24 hours with and without UV-illumination. The solutions were subsequently centrifuged to remove the powder, and the Bi and V contents of the supernatant were analyzed using inductively-coupled plasma optical emission spectroscopy (ICP-OES).

RESULTS AND DISCUSSION

Figure 2a shows a photograph of a 100 nm $BiVO_4$ film on FTO substrate (left) next to a bare FTO substrate (right). The yellow color indicates the presence of $BiVO_4$, and only a small amount of light scattering is shown by the sample. The microstructure of the film is characterized using planar and cross-sectional SEM, as shown in Fig. 2b and c, respectively. A typical particle size of ~100 nm is obtained, which is much smaller than that of earlier samples prepared by less well-defined glass spray nozzles [9].

Photochemical stability of $BiVO_4$

The stability of $BiVO_4$ in solutions of different pH, with and without illumination, is shown in Figure 2d.The material is found to be stable between pH 3 and 11, both in the dark and under UV illumination. Under both acidic (pH < 3) and alkaline (pH > 11) conditions the vanadium dissolves preferentially, exceeding the amount of dissolved bismuth by a factor of ~100. At pH 1 the dissolution rate is much higher under illumination than in the dark, whereas the dark- and photo-corrosion rates are approximately equal under alkaline conditions.

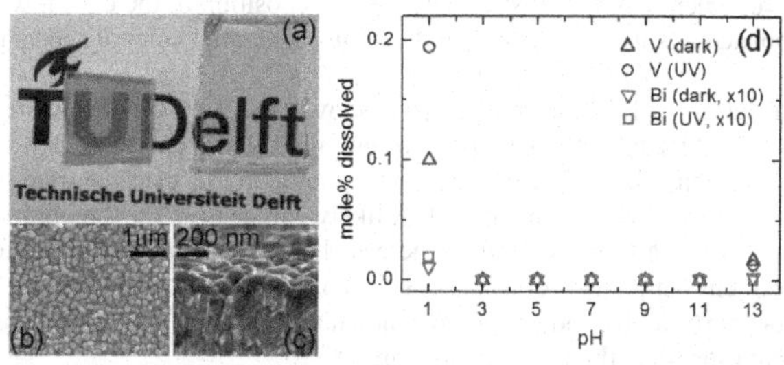

Figure 2. (a) Photograph of 100 nm thick $BiVO_4$ on an FTO-substrate (left) and a bare FTO-substrate (right). (b) Planar, and (c) cross-section scanning electron microscopy image of 100 nm thick $BiVO_4$ sample. (d) Mole% of Bi (×10) and V dissolved after 24 hours in the dark and under UV illumination in aqueous KOH/HCl solutions of varying pH.

Photoelectrochemical characterization of thin film $BiVO_4$

Figure 3 shows the photocurrent-voltage curves of 200 nm-thick uncatalyzed and Co-Pi catalyzed $BiVO_4$ samples under chopped AM1.5 illumination. The $BiVO_4$ was illuminated via the substrate, i.e., the back-side. The uncatalyzed sample (black curve) shows a photocurrent of ~0.7 mA/cm^2 at 1.23 V vs. RHE. While this is higher than previous reported values for bare $BiVO_4$ [2-8], it is still a factor of 10 lower than the theoretical maximum photocurrent that can be achieved by $BiVO_4$ (7.5 mA/cm^2, assuming that all photons with energies >2.4 eV are absorbed). Based on previous reports on $BiVO_4$, slow water oxidation kinetics is the performance limiting factor at this illumination intensity [4-8]. In order to overcome this, 30 nm Co-Pi catalyst is deposited on the surface of $BiVO_4$ photoanode. As a result, a photocurrent of ~1.8 mA/cm^2 at 1.23 V vs. RHE is achieved, as shown in Fig. 3 (red curve). This significant improvement illustrates the improved catalytic activity of the photoanode upon the application of a Co-Pi catalyst.

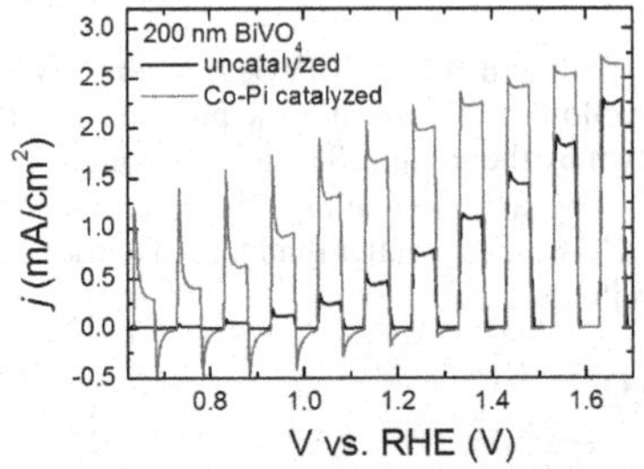

Figure 3. Chopped photocurrent-voltage measurement of uncatalyzed and Co-Pi catalyzed $BiVO_4$ of 200 nm thick under AM1.5 back-side illumination. The scan rate is 10 mV/s.

Despite the large enhancement of the catalytic activity for water oxidation, the red curve in Fig. 3 still shows pronounced photocurrent transients at potentials below ~1.2 V_{RHE}. We attribute this to sub-optimal coverage of the $BiVO_4$ with Co-Pi during the electrodeposition process. Further improvements may be possible by *photo*-deposition of the Co-Pi, which ensures that the Co-Pi is deposited at surface sites where the photo-generated holes are most readily available [12].

The photocurrent of Co-Pi catalyzed $BiVO_4$ is shown as a function of thickness in Figure 4a. A 300 nm $BiVO_4$ film gives the highest photocurrent of ~1.9 mA/cm^2 at 1.23 V vs. RHE under back-side illumination. This is one of the highest photocurrents ever reported for $BiVO_4$ catalyzed with a low-cost earth-abundant catalyst. A likely explanation for the optimum thickness is that thinner films absorb less light, whereas charge transport is limiting for thicker films. This is in agreement with a theoretical calculation of the optical absorption of $BiVO_4$ photoanodes based on its refractive index [13]. As shown in Figure 4b, the optical absorption remains nearly constant for films thicker than 300 nm.

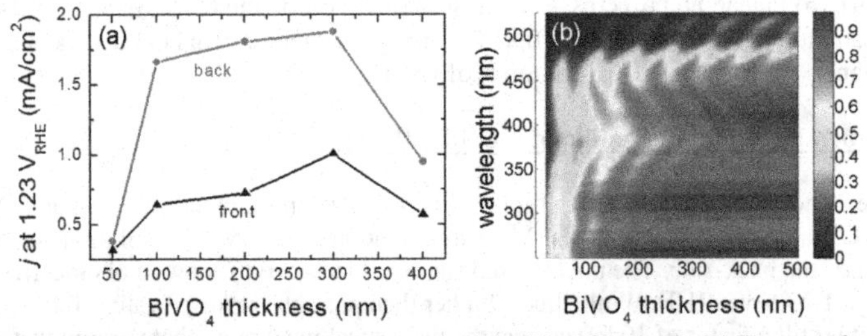

Figure 4. (a) AM1.5 photocurrent of $BiVO_4$ catalyzed with a 30 nm Co-Pi film as a function of the $BiVO_4$ thickness under front- and back-side illumination. (b) 2D color plot of the calculated absorption as a function of wavelength and $BiVO_4$ film thickness.

To further improve the performance of the films, it is important to first determine the performance-limiting factor of the Co-Pi catalyzed $BiVO_4$ photoanode. An elegant way to analyze the performance of a photoelectrode is recently reported by Dotan et al [14]. The total photocurrent due to water splitting can be described by the following equation:

$$J_{H_2O} = J_{abs} \times \eta_{sep} \times \eta_{ox} \tag{1}$$

where J_{abs} is the photon absorption rate expressed as a current, η_{sep} is the charge separation efficiency, and η_{ox} is the water oxidation efficiency. When hydrogen peroxide is added into the electrolyte solution, it acts as an effective hole scavenger. This results in a 100% oxidation efficiency ($\eta_{ox} = 1$), and the following equation can be written:

$$J_{H_2O_2} = J_{abs} \times \eta_{sep} \tag{2}$$

Based on equation (1) and (2), the charge separation efficiency and the oxidation efficiency can be obtained from the following equations:

$$\eta_{sep} = \frac{J_{H_2O_2}}{J_{abs}} \quad (3); \qquad \eta_{ox} = \frac{J_{H_2O}}{J_{H_2O_2}} \quad (4)$$

Figure 5a shows photocurrent-voltage curves of 100 nm thick $BiVO_4$ sample in a normal electrolyte (black curve) and in the presence of H_2O_2 (blue curve). The photocurrent of a Co-Pi catalyzed sample is also shown (red curve). Based on these curves, the oxidation efficiency and the charge separation efficiency of the samples have been calculated using Eqs. (3) and (4) and plotted in Figure 5b. The Co-Pi clearly enhances the oxidation efficiency of $BiVO_4$, reaching values >90% at potentials positive of 1.2 V_{RHE}.

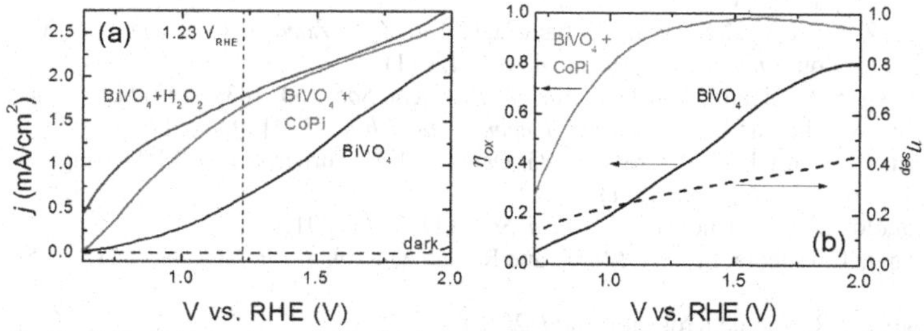

Figure 5. (a) AM1.5 photocurrent vs. voltage for a 100 nm $BiVO_4$ photoelectrode (with and without H_2O_2 in the electrolyte) and a Co-Pi catalyzed $BiVO_4$ photoelectrode at a scan rate of 50 mV/s. The dark curve is shown by the horizontal black dashed line. (b) Oxidation and charge separation efficiencies of uncatalyzed and Co-Pi catalyzed $BiVO_4$ as a function of applied bias.

Fig. 5b shows that the charge separation efficiency is the main problem to be addressed. About 60-80% of the electron-hole pairs recombine before reaching the interfaces. A pronounced difference in the photocurrents under front- and back-side illumination reported in our previous studies indicates that slow electron transport is responsible for this [9,10]. Future efforts to improve the performance of $BiVO_4$ should therefore be focused to solve this limitation. Possible strategies to achieve this are doping and/or the use of guest-host nanostructures [15] in combination with plasmonic- or resonance-enhanced optical absorption [16].

CONCLUSIONS

We have shown that high quality $BiVO_4$ films can be made with a low-cost and scalable spray pyrolysis method. The material is photochemically stable between pH 3 and 11, and an AM1.5 photocurrent of 1.9 mA/cm^2 at 1.23 V vs. RHE is achieved using a 300 nm thick $BiVO_4$ film modified with a 30 nm Co-Pi catalyst layer. The photocurrent is limited by a low charge separation efficiency caused by the inherently slow electron transport in $BiVO_4$. Further improvement of $BiVO_4$ performance therefore requires solving this limitation.

ACKNOWLEDGMENTS

We gratefully acknowledge European Comission's Framework Project 7 (NanoPEC, Project 227179) for the financial support of this work. We thank Maurin Cornuz (EPFL) for the cross-section SEM imaging, Maarten de Nier for the picture of the deposition setup, and Lennard Mooij for the stability measurements.

REFERENCES

1. S. Tokunaga, H. Kato, and A. Kudo, *Chem. Mater.* **13**, 4624 (2001).
2. P. Chatchai, Y. Murakami, S. Y. Kishioka, A. Y. Nosaka, and Y. Nosaka, *Electrochem. Solid State Lett.* **11**, H160 (2008).
3. K. Sayama, N. Wang, Y. Miseki, H. Kusama, N. Onozawa-Komatsuzaki, and H. Sugihara, *Chem. Lett.* **39**, 17 (2010).
4. W. J. Luo, Z. S. Yang, Z. S. Li, J. Y. Zhang, J. G. Liu, Z. Y. Zhao, Z. Q. Wang, S. C. Yan, T. Yu, and Z. G. Zou, *Energy Environ. Sci.* **4**, 4046 (2011).
5. D. K. Zhong, S. Choi, and D. R. Gamelin, *J. Am. Chem. Soc.* **133**, 18370 (2011).
6. T. H. Jeon, W. Choi, and H. Park, *Phys. Chem. Chem. Phys.* **13**, 21392 (2011).
7. S. K. Pilli, T. E. Furtak, L. D. Brown, T. G. Deutsch, J. A. Turner, and A. M. Herring, *Energy Environ. Sci.* **4**, 5028 (2011).
8. J. A. Seabold and K. S. Choi, *J. Am. Chem. Soc.* **134**, 2186 (2012).
9. Y. Q. Liang, T. Tsubota, L. P. A. Mooij, and R. Van de Krol, *J. Phys. Chem. C* **115**, 17594 (2011).
10. F. F. Abdi and R. Van de Krol, *submitted* (2012).
11. M. W. Kanan and D. G. Nocera, *Science* **321**, 1072 (2008).
12. E. M. P. Steinmiller and K. S. Choi, *Proc. Nat. Acad. Sci. USA* **106**, 20633 (2009).
13. Z. Y. Zhao, Z. S. Li, and Z. G. Zou, *Phys. Chem. Chem. Phys.* **13**, 4746 (2011).
14. H. Dotan, K. Sivula, M. Gratzel, A. Rothschild, and S. C. Warren, *Energy Environ. Sci.* **4**, 958 (2011).
15. K. Sivula, F. Le Formal, and M. Graetzel, *Chem. Mater.* **21**, 2862 (2009).
16. S. Linic, P. Christopher, and D. B. Ingram, *Nature Mater.* **10**, 911 (2011).

Mater. Res. Soc. Symp. Proc. Vol. 1446 © 2012 Materials Research Society
DOI: 10.1557/opl.2012.870

Artificial Photosynthesis - Use of a Ferroelectric Photocatalyst

Steve Dunn [a] **and Matt Stock**[b]

[a] School and Engineering and Materials, Queen Mary University of London, E1 4NS, UK

[b] Cranfield University, Cranfield, MK43 0AL

Email: s.c.dunn@qmul.ac.uk

ABSTRACT

The solid-gas phase photoassisted reduction of carbon dioxide (artificial photosynthesis) was performed using ferroelectric lithium niobate and titanium dioxide as photocatalysts. Illumination with a high pressure mercury lamp and visible sunlight showed lithium niobate achieved unexpectedly high conversion of CO_2 to products despite the low levels of band gap light available and outperformed titanium dioxide under the conditions used. The high reaction efficiency of lithium niobate is explained due to its strong remnant polarization (70 $\mu C/cm^2$) thought to allow longer lifetime of photo induced carriers as well as an alternative reaction pathway.

INTRODUCTION

The threat of climate change due to rising levels of green house gases from human activity has made the need for renewable sources of energy a priority. One promising area of providing green energy is artificial photosynthesis[1], a process that mimics mechanisms of nature. A semiconductor can be used as a photocatalyst absorbing light and using the energy to chemically convert CO_2 and water into fuel stock in a carbon neutral mechanism. In this work we describe artificial photosynthesis taking place on ferroelectric lithium niobate ($LiNbO_3$) producing formic acid and formaldehyde at efficiencies exceeding those of other unmodified metal oxide photocatalysts such as titanium dioxide (TiO_2).

Artificial photosynthesis has been investigated using a variety of approaches[2] including replication of the chemical reactions that take place in plants[3] and concerted efforts using ruthenium complexes and structures as a dye[4] to sensitise wide band gap semiconductors to visible light. Metal oxide semiconductor systems including WO_3, Fe_2O_3 and ZnO have also proven effective as photoactive surfaces for water purification[5], alternative electron transport materials in nanostructured photovoltaics[6] and as a means to sensitise titanium dioxide to longer wavelength radiation through doping.

One aspect of surface photochemistry that has not been extensively addressed is to use materials that sustain a dipole to separate the photoinduced electrons and holes - ferroelectric materials. Early work by [7] showed selective oxidation and reduction reactions take place on $BaTiO_3$. Subsequent work on other ferroelectric systems $PbZr_{0.3}Ti_{0.7}O_3$[8; 9] and $LiNbO_3$[10] indicated that the dipole in the ferroelectric material determines the space charge layer structure. The surface charge has also been shown to interact with the species producing a tightly bound double layer[11]. There has been discussion that such a tightly bound layer can alter the nature of bonding in physisorbed materials[12]. $LiNbO_3$ was selected due to remnant polarization of 70 μC/cm^2 [13] as compared to 30 μC/cm^2 [14] ($KNbO_3$) and 25 μC/cm^2 [15] (PZT). $LiNbO_3$ has been previously tested in solid-liquid reactions and showed a low reaction efficiency[16]. Here we use a solid-gas reaction scheme. The remnant polarization of the ferroelectric material was exploited to achieve a conversion of CO_2 into hydrocarbon despite a wide band gap of 3.78 eV[17]. Our results show polar-semiconductors are exciting new materials to drive artificial photosynthesis.

EXPERIMENTAL METHODS

High purity single crystal $LiNbO_3$ was purchased from MTI Corporation[18] and ground using a mortar and pestle to give a powder with particles ranging from 100nm to 5μm. Anatase TiO_2 325 mesh powder with an average particle size of 200nm was purchased from Sigma Aldrich and used as provided. A reaction vessel was constructed from a PTFE beaker, quartz lid to allow transmission of all incident light, needle valve to sample the contents and two pieces of Tufnol sheet fitted with threaded bar to allow the lid to be compressed to form a gas tight seal.

The vessel was prepared by loading the beaker with the PTFE holder containing either $LiNbO_3$ or TiO_2 powder with an exposed flat circular surface of 0.126 cm^2. A reservoir of 5ml of distilled water was created before filling with CO_2 gas to create a 30:70 mix with air. Irradiation was carried out either by a UV Honle Hg lamp for 6 hours or by natural sun light for 15 hours. After irradiation the vessel was allowed to cool and samples were collected from the water reservoir for analysis. A 400W Honle lamp fitted with an Hg doped bulb (250-400nm) was used to carry out all UV irradiation. The output was measured and found to be 28.56mW/cm^2. Experiments under natural sunlight were carried out under AM1.5 illumination. Liquid samples (2.5ml) were collected and analysed using gas chromatography/mass spectroscopy (GC/MS). Before analysis samples were first prepared using a chemical mediation method[19]. GC/MS was performed using an Agilent Model 6890N fitted with a Model 5973N quadrapole mass selective detector, a Model 7694 headspace autosampling unit and a Phenomenex ZB-Wax column 30m long with a 0.32mm i.d. and 0.5μm film thickness. Energy conversion efficiency (ECE) was calculated using the available photoenergy as defined by the band gap of the semi-conductor and emission spectrum of the bulb and energy stored in hydrocarbon products (combustion product).

RESULTS AND DISCUSSION

Figure 1 outlines the key steps of artificial photosynthesis. First absorption of light by the photocatalyst excites electrons from the valence band to the conduction band and simultaneously creates valence band holes. The conduction band electrons can then drive reduction of CO_2 to CO_2^- and the valence band holes oxidize water to H^+. The reactive species CO_2^- and H^+ then form products including formic acid and formaldehyde.

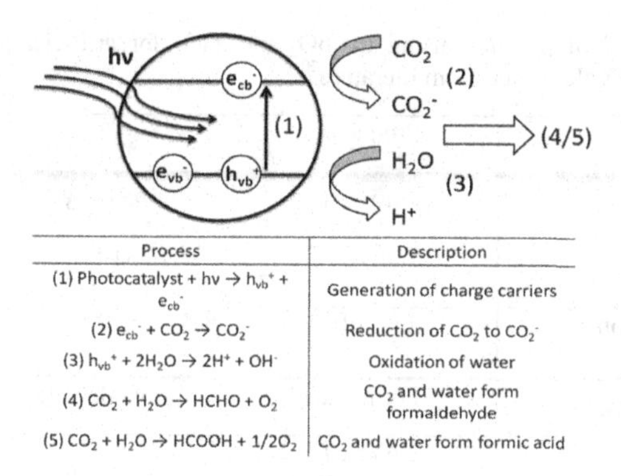

Process	Description
(1) Photocatalyst + hv → h_{vb}^+ + e_{cb}^-	Generation of charge carriers
(2) e_{cb}^- + CO_2 → CO_2^-	Reduction of CO_2 to CO_2^-
(3) h_{vb}^+ + $2H_2O$ → $2H^+$ + OH^-	Oxidation of water
(4) CO_2 + H_2O → HCHO + O_2	CO_2 and water form formaldehyde
(5) CO_2 + H_2O → HCOOH + $1/2O_2$	CO_2 and water form formic acid

Figure 1, The principal steps of artificial photosynthesis.

$LiNbO_3$ and TiO_2 were compared as photocatalysts for artificial photosynthesis under both UV and visible light. Samples collected were analysed using GC/MS to determine the levels of formic acid and formaldehyde present. The results comparing $LiNbO_3$ and TiO_2 are shown in Table 1 and Table 2. Blank runs were carried out in the absence of a photocatalyst and showed neither formic acid or formaldehyde was present under any radiation. In all cases the experiments for UV illumination with $LiNbO_3$ and TiO_2 were conducted and analyzed under identical conditions. The experiment for the visible light reaction of $LiNbO_3$ was carried out under AM1.5 while the data for TiO_2 taken from the literature.

Table 1, Yield of products using $LiNbO_3$ and TiO_2 for artificial photosynthesis under UV irradiation.

UV irradiation	Products		
$LiNbO_3$	Formic acid	Formaldehyde	Total
Mol m^2 h^{-1}	1.24 x10^{-2}	7.07 x10^{-4}	1.31 x10^{-2}
l m^2 h^{-1} of CO_2 converted	2.78 x10^{-1}	1.58 x10^{-2}	2.94 x10^{-1}
ECE (%)			0.34
TiO_2	Formic acid	Formaldehyde	Total
Mol m^2 h^{-1}	4.15 x10^{-3}	1.77 x10^{-4}	4.33 x10^{-3}
l m^2 h^{-1} of CO_2 converted	9.30 x10^{-2}	3.96 x10^{-3}	9.70 x10^{-2}
ECE (%)			0.05

Table 2, Yield of products using $LiNbO_3$ and TiO_2 for artificial photosynthesis under visible irradiation. *Value taken from literature[20]

Visible irradiation	Products		
$LiNbO_3$	Formic acid	Formaldehyde	Total
Mol m^2 h^{-1}	9.26 x10^{-4}	3.98 x10^{-5}	9.65 x10^{-4}
l m^2 h^{-1} of CO_2 converted	2.07 x10^{-2}	1.31 x10^{-2}	2.16 x10^{-2}
TiO_2*	Formic acid	Formaldehyde	Total
Mol m^2 h^{-1}	2.63 x10^{-6}	-	2.63 x10^{-6}
l m^2 h^{-1} of CO_2 converted	5.88 x10^{-5}	-	5.88 x10^{-5}

The results in Table 1 show under UV irradiation $LiNbO_3$ produced 1.31x10^{-2} mol m^2 h^{-1} of products. This is 3 times that when using TiO_2 of 4.33x10^{-3} mol m^2 h^{-1}. It was also found under these conditions $LiNbO_3$ achieved an energy conversion efficiency (ECE) of incident light to chemical energy of 0.34%, 7 times that of 0.05% when using TiO_2.

Table 2 shows under visible irradiation that $LiNbO_3$ produced 9.65x10^{-4} mol m^2 h^{-1} of products, 36 times that of 2.63x10^{-6} mol m^2 h^{-1} produced by TiO_2. These results are surprising as it would be expected that TiO_2 should produce more products as it has been shown to be an effective photocatalyst. There are a number of reasons why $LiNbO_3$ might be generating such results. The first is that the remnant polarization, not found in semiconductors such as TiO_2, is affecting the artificial photosynthesis process and improving reaction efficiency. It is also possible that the reaction pathway available for artificial photosynthesis from $LiNbO_3$ is also a more energetically favourable.

When electrons and holes are generated some recombine as internal or surface recombination. The loss of carriers reduces the photocatalyst's efficiency. In ferroelectric materials, such as $LiNbO_3$, the remnant polarization produces an electric field – this is similar to the electric field found in a typical p-n junction. This field separates photo generated electrons and holes as show in Figure 2. Experimental evidence for longer lifetimes of carriers has been provided as the decay time of polaron photoluminescence in $LiNbO_3$[21] is very high (9 μs) suggesting the recombination is inhibited.

$LiNbO_3$'s remnant polarization also leads to a charge experienced at the interface and interacts with species in contact with the surface producing a tightly bound layer[11]. We propose that in earlier work focusing on liquid-solid catalysts it is precisely this tightly bound layer that prevented the system from producing product. In the gas-sold reaction there is significantly more CO_2 available on the surface of the $LiNbO_3$. This bound CO_2 can react with photoexcited holes and electrons in the $LiNbO_3$. One such species that can be formed is the CO_2^-. It has been shown the physisorption of molecules on ferroelectric surfaces[11] affects the localisation of electrons in the molecule as well as the bond angles. When performing artificial photosynthesis $LiNbO_3$ can absorb linear CO_2 on the C+ face as in Figure 3[12]. The $LiNbO_3$ is unable to directly oxidize water due to the valance band position and so we propose that a super oxide anion is formed and performs the oxidation processes. Figure 4 compares the potential of the valence and conduction bands of $LiNbO_3$ and TiO_2 in relation to the formation of formic acid and formaldehyde. Valence band holes can oxidize water but conduction band electrons cannot reduce CO_2 to CO_2^{-}. $LiNbO_3$ can drive the reaction of CO_2 and water. $LiNbO_3$'s conduction band electrons are highly reducing and can reduce CO_2 to CO_2^{-} or CO_2^{2-} unlike TiO_2. This reaction step can allow a different reaction path way using $LiNbO_3$ due to its highly reducing conduction band electrons. These highly reactive species are available to react with the bound water and promote new reactive species to be generated.

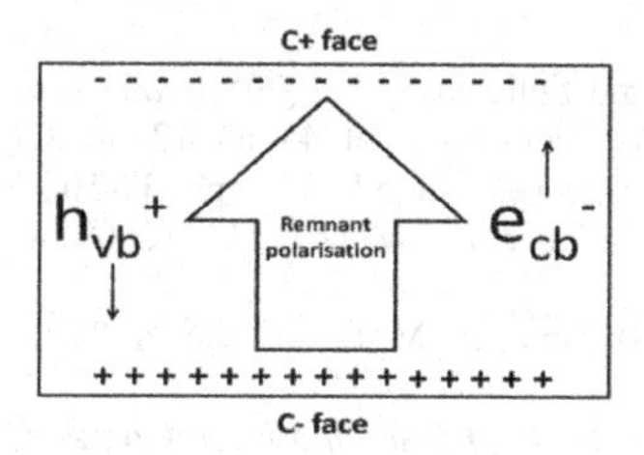

Figure 2, Separation of photogenerated carriers in a ferroelectric.

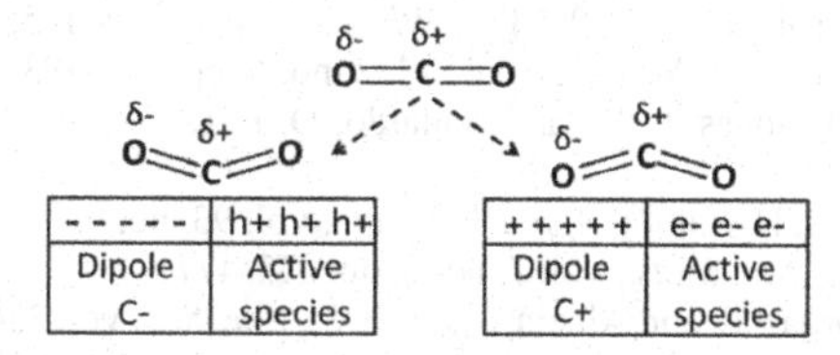

Figure 3, Adsorption of CO_2 onto the C+ and C- faces of $LiNbO_3$ leading to bending.

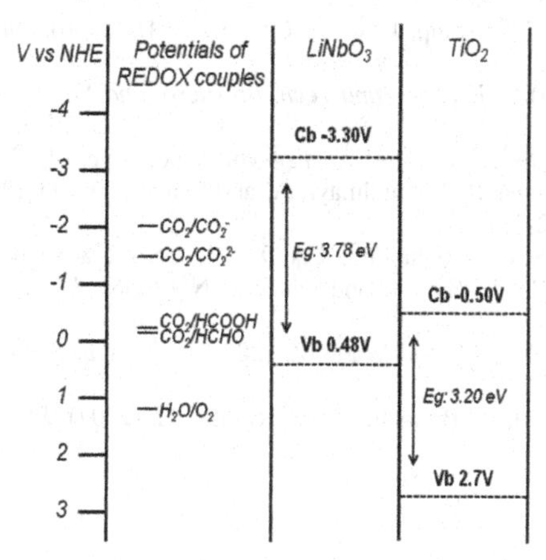

Figure 4, Band positions of $LiNbO_3$ and TiO_2 Vs NHE in relation to the REDOX reactions.

CONCLUSION

Under UV irradiation $LiNbO_3$ was found to produce 7 times the mols m^{-2} $hour^{-1}$ than when using TiO_2 and under visible irradiation $LiNbO_3$ formed 36 times the mols m^{-2} $hour^{-1}$ than when using TiO_2. The polarisation of $LiNbO_3$ facilitates improved carrier life times as well as altered adsorption of reactants due to surface charge leading to easier injection of charge carriers both of which help improve $LiNbO_3$'s efficency.

REFERENCES

1. Hashimoto, K., Irie, H. and Fujishima, A. (2005), *Jap. J. of App. Phys., Part 1: Regular Papers and Short Notes and Review Papers,* vol. 44, no. 12, pp. 8269-8285.
2. Bolton, J. R. (1996), *Solar Energy,* vol. 57, no. 1, pp. 37-50.
3. Ghirardi, M. L., Dubini, A., Yu, J. and Maness, P. -. (2009), *Chem. Soc. Rev.,* vol. 38, no. 1, pp. 52-61.
4. Alstrum-Acevedo, J. H., Brennaman, M. K. and Meyer, T. J. (2005), *Inorg. Chem.,* vol. 44, no. 20, pp. 6802-6827.
5. Serpone, N., et al. (1995), *J. Photochem. and Photobio., A: Chem.,* vol. 85, no. 3, pp. 247-255.
6. Lévy-Clément, C., et al. (2005), *Ad. Mat.,* vol. 17, no. 12, pp. 1512-1515.
7. Giocondi, J. L. and Rohrer, G. S. (2001), *J. Phys. Chem. B,* vol. 105, no. 35, pp. 8275-8277.
8. Kalinin, S. V., et al. (2002), *Nano Letters,* vol. 2, no. 6, pp. 589-593.
9. Dunn, S., Tiwari, D., Jones, P. M. and Gallardo, D. E. (2007), *J. Mats Chem.,* vol. 17, no. 42, pp. 4460-4463.
10. Dunn, S. and Tiwari, D. (2008), *App. Phys. Lett.,* vol. 93, no. 9.
11. Li, D., et al. (2008), *Nat. Mats.,* vol. 7, no. 6, pp. 473-477.
12. Cabrera, A. L., Vargas, F. and Albers, J. J. (1995), *Sur. Sci.,* vol. 336, no. 3, pp. 280-286.
13. Yang, W., Rodriguez, B. J., Gruverman, A. and Nemanich, R. J. (2004), *App. Phys. Letts,* vol. 85, no. 12, pp. 2316-2318.
14. Ramos-Moore, E., Baier-Saip, J. A. and Cabrera, A. L. (2006), *Sur. Sci.,* vol. 600, no. 17, pp. 3472-3476.
15. Yamada, H. (1999), *J. Vac. Sci. and Tech. B: Micro. and Nano. Struc.,* vol. 17, no. 5, pp. 1930-1934.
16. Ulman, M., et al. (1982), *Int. J Sol. Energy,* vol. 1, no. 3, pp. 213-222.
17. Thierfelder, C., Sanna, S., Schindlmayr, A. and Schmidt, W. G. (2010), *Phys. Stat. Solidi (C),* vol. 7, no. 2, pp. 362-365.
18. 860 South 19th Street, Richmond, CA 94804, USA , available at: www.mtixtl.com.
19. del Barrio, M. -., Hu, J., Zhou, P. and Cauchon, N. (2006), *J. Pharma. and Bio. Anal.,* vol. 41, no. 3, pp. 738-743.
20. Zhao, Z., Fan, J., Xie, M. and Wang, Z. (2009), *J. Cleaner Prod.,* vol. 17, no. 11, pp. 1025-1029.
21. Harhira, A., Guilbert, L., Bourson, P. and Rinnert, H. (2007), *Phys. Stat. Solidi (C) s,* vol. 4, no. 3, pp. 926-929.

Mater. Res. Soc. Symp. Proc. Vol. 1446 © 2012 Materials Research Society
DOI: 10.1557/opl.2012.952

Copper Tungstate ($CuWO_4$)–Based Materials for Photoelectrochemical Hydrogen Production

Nicolas Gaillard[1,*], Yuancheng Chang[1], Artur Braun[1,2] and Alexander DeAngelis[1]

[1]Hawaii Natural Energy Institute, University of Hawaii at Manoa, Honolulu, HI 96822, USA

[2]Laboratory for High Performance Ceramics, Empa, Swiss Federal Laboratories for Materials Science and Technology, Überlandstrasse 129, CH - 8600 Dübendorf, Switzerland

(*) Corresponding Author
Email: ngaillar@hawaii.edu
Phone: 1.808.956.2342

ABSTRACT

We report in this communication on the photoelectrochemical (PEC) performances of copper tungstate ($CuWO_4$) material class. This study was performed on 2-micron thick samples fabricated using a low-cost co-sputtering deposition process, followed by an 8-hour long annealing at 500°C in argon. Microstructural analysis pointed out that the post-deposition treatment was critical to achieve photocatalytic activity. Subsequent characterizations revealed that polycrystalline $CuWO_4$ photoanodes owned promising characteristics for solar-assisted water splitting, i.e (i) an optical band-gap of 2.2 eV, (ii) a flat-band potential of -0.35 V vs. SCE and (iii) conduction and valence band-edges that straddle water splitting redox potentials. $CuWO_4$ photoanodes generated 400 $\mu A.cm^{-2}$ at 1.6V vs. SCE under simulated $AM1.5_G$ illumination in 0.33M H_3PO_4 with virtually no dark current up to this potential. Impedance analysis pointed out that large charge transfer resistances (2,500 $\Omega.cm^2$) could be the main weakness of this material class. Current research activity is focused on solving this issue to achieve higher PEC performances.

INTRODUCTION

Three decades after the demonstration of photoelectrochemical (PEC) water splitting by Fujishima and Honda with TiO_2 [1], intensive research is still ongoing to identify a suitable semiconductor to be integrated in an efficient, cost effective, and reliable PEC system. Among all candidates, TiO_2 and WO_3 are still drawing lots of attention as they offer good resistance to corrosion and are inexpensive to produce. One of the major issues to solve remains their large band-gap that restrains solar-to-hydrogen efficiency to only few percent [2]. Numerous attempts have been done to reduce the band-gap of metal oxides, mainly via incorporation of foreign element such as nitrogen [3]. Unfortunately, this method usually leads to an increase in structural defects and poor PEC performances [4]. From this point of view, it appears that the best research strategy might be to focus on metal oxides inherently having appropriate absorption properties and improve, if necessary, their transport and/or catalytic properties. With an optical band gap of 2.2 eV, copper tungstate ($CuWO_4$) should be considered as a potential candidate for PEC hydrogen production.

As of today, most publications on $CuWO_4$ report on its fundamental properties but only a handful on its potential as a photoactive system. In the present communication, we report on the synthesis of $CuWO_4$ thin film via a low-cost process for PEC hydrogen production. The impact of post-deposition annealing on the material microstructure and PEC performances is discussed. The surface energetics of $CuWO_4$ obtained with electrochemical impedance spectroscopy are presented and compared with those of WO_3. Finally, $CuWO_4$ charge transport resistance is analyzed and discussed.

EXPERIMENTAL DETAILS

Copper tungstate thin films were deposited by reactive RF magnetron co-sputtering method using metallic Cu (99.99%) and W (99.95%) targets. The substrates (SnO_2:F-coated glass, TEC 15 Hartford Glass) were cleaned in a soap/DI-water/acetone/DI-water/methanol/DI-water/isopropanol sequence and placed in the deposition chamber. After reaching a base pressure of $2x10^{-6}$ Torr the substrates were heated to ~275°C. Argon (Ar) and oxygen (O_2) were then introduced into the chamber with gas flows set for an $[O_2]/([Ar]+[O_2])$ ratio of 36% (working pressure: 9.0 mTorr). The copper-to-tungsten ratio in the films was controlled by adjusting the RF power applied to each target. Post-deposition thermal treatment was performed in argon at 500°C for 8 hours in a ceramic tube furnace (Micropyretics Heater International, Inc.).

X-ray diffraction analysis was performed with a Rigaku Miniflex II X-ray diffractometer using a Cu cathode (K_α 1.546 Å) powered at 20kV and 20mA. The reflectance and transmission of annealed $CuWO_4$ films were measured with a Lambda 2 Perkin Elmer spectrophotometer between 350 and 1100 nm. Electrochemical analyses were performed with a Gamry 600 potentiostat. The test cell consisted of a $CuWO_4$ PEC electrode (typically 1.5 cm^2), a platinum foil counter electrode (10 cm^2) and an SCE reference electrode. Aqueous 0.33M H_3PO_4 electrolytes (pH 1.35) were used for all tests. Photoanodes were irradiated with simulated Air Mass 1.5 Global ($AM1.5_G$) illumination provided by an Oriel class A solar simulator equipped with a xenon bulb and a $AM1.5_G$ filter. The irradiance of the simulator was calibrated using an International Light Technologies 900 spectrophotoradiometer. This calibration was done such that the Xe bulb output power within $CuWO_4$ absorption range (up to 560 nm wavelength) matches with the $AM1.5_G$ radiance in the same range (Figure 1). Electrochemical impedance spectroscopy (EIS) was performed at various potentials and frequencies (1 Hz to 1 MHz) to identify the $CuWO_4$ flat-band potential and the carrier concentration. Finally, linear sweep voltammetry was measured in 3-electrode configuration.

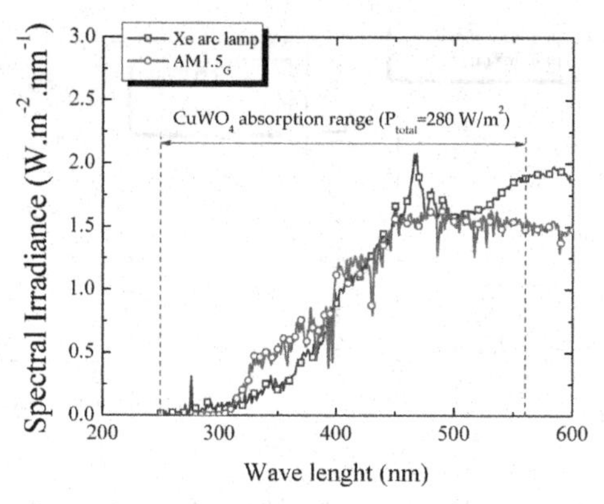

Fig. 1. Spectral irradiance of simulated $AM1.5_G$ illumination measured with an ITL900 spectroradiometer. The output power of the Xe arc lamp is set such that its irradiance corresponded to the one of $AM1.5_G$ within $CuWO_4$ absorption range (280 $W.m^{-2}$ up to 560 nm wavelength).

RESULTS AND DISCUSSION

Figure 2.a presents the X-ray diffractograms acquired on as-deposited and annealed copper tungstate thin films (2 micron thick). One can observe that the temperature used during the film synthesis (275°C) leads essentially to an amorphous material, as only SnO_2:F diffraction peaks were observed. This observation is consistent with the one reported recently by Chen [5]. When a post-deposition annealing was performed at 500°C, a series of intense peaks were observed and identified as wolframite $CuWO_4$ (PDF 21-0307). A rapid assessment of both amorphous and polycrystalline $CuWO_4$ PEC properties revealed that as-deposited (amorphous) films did not exhibit any response to incident $AM1.5_G$ irradiation (as reported also by Chen [5]), whereas those annealed were photoactive. Therefore, research efforts were directed toward polycrystalline copper tungstate obtained by post-annealing in argon at 500°C.

The Tauc plot of annealed $CuWO_4$ film derived from reflection and transmission spectra (assuming indirect optical transition [6]) is presented in Figure 2.b. Here, an optical band-gap of 2.25 (±0.05) eV was obtained, in good agreement with values reported for this material class[7]. When compared to other metal oxides such as tungsten oxide (E_G = 2.6 eV[8,9]), the absorption properties measured on $CuWO_4$ in the present study are more suitable for practical PEC applications, in which a band gap value of approximately 2.0 eV is considered as ideal[10,11]. Also presented in Figure 2.b is the Tauc plot calculated from photo-spectroscopy measurements. With this method, a comparable band-gap value (i.e. 2.2 (±0.05) eV) was obtained. This indicates that photons with energy equal or greater than 2.2 eV are effectively converted to electrons and hole by $CuWO_4$ and participate to the water splitting process.

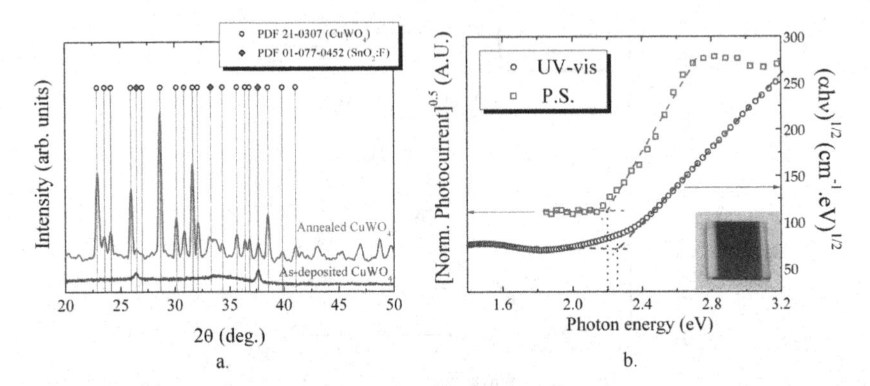

Fig. 2. (a) X-ray diffractograms acquired on as-deposited (blue) and annealed (red) $CuWO_4$ thin films and (b) Tauc plots of an annealed $CuWO_4$ thin film derived from optical (UV/visible) and photocurrent spectroscopy methods. (Inset) Picture of a brownish $CuWO_4$ sample (1"×1").

EIS analyses were performed in 0.33M H_3PO_4 under AM1.5$_G$ illumination to assess the conductivity type and donor concentration of polycrystalline $CuWO_4$ thin films. Capacitance values were extracted from EIS spectra (fitted using a constant phase element equivalent circuit) and plotted in a Mott-Schottky (MS) fashion (Figure 3.a). The positive slope of the MS plot clearly shows that polycrystalline $CuWO_4$ thin films owned n-type conductivity. The MS plot also indicates a flat-band potential (V_{FB}) of -0.35 V vs. SCE for $CuWO_4$. When compared to reactively sputtered n-type WO_3 (0.15 V vs. SCE [2]), this value corresponds to a cathodic shift of 500 mV. In other words, $CuWO_4$ material would generate photocurrent at a potential half a volt lower than WO_3. Finally, donor concentration was calculated from the MS slope, assuming a dielectric constant for $CuWO_4$ of 83. Here N_D was estimated to be 1.20×10^{21} cm^{-3}, a value two orders of magnitude higher than that reported for monocrystalline $CuWO_4$ (5.50×10^{19} cm^{-3}) [6] but comparable to that recently reported by others for polycrystalline $CuWO_4$ thin films (2.7×10^{21} cm^{-3}) [7].

Data gathered from Tauc and MS plots were combined to depict the surface energetics of polycrystalline $CuWO_4$. For the sake of comparison, identical characterizations were performed on reactively sputtered WO_3. The results of this comparison, presented in Figure 3.b, evidenced that between the two oxides $CuWO_4$ owns the most favorable surface energetics for solar-assisted water splitting. In fact, both $CuWO_4$ conduction band minimum (CBM) and valence band maximum (VBM) straddle the hydrogen and oxygen evolution reduction potentials, respectively. More importantly, the VBM in $CuWO_4$ is located 700 mV closer to H_2O/O_2 redox potential than that of WO_3, an ideal condition previously attempted to be achieved with incorporation of nitrogen p states near WO_3 VBM [3,4]. Recent theoretical studies[12] and experimental analyses[7] have shown that the shift of VBM in $CuWO_4$ when compared to WO_3 can be directly attributed to the presence of Cu(3d) states. The raising of the CBM reported in the present communication report is currently not supported by any theoretical study, although some reports have shown than Cu(3d) might also contribute to CBM[13]. It is worth mentioning that the position of CBM reported here is consistent with other experimental results[14]. Additional characterization will be required to clarify this point.

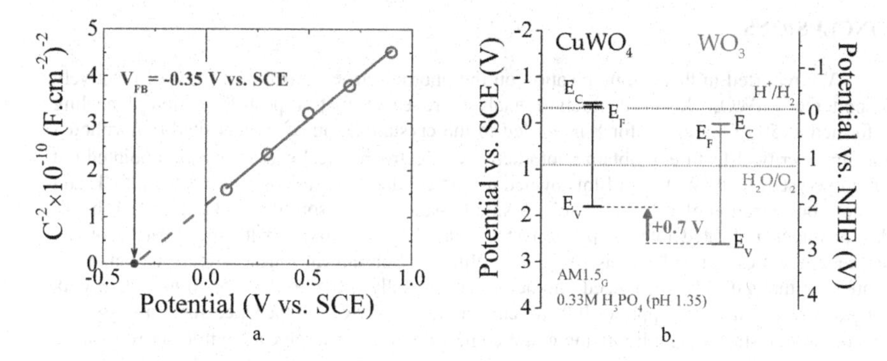

Fig. 3. (a) Mott-Schottky plot obtained on annealed $CuWO_4$ thin films and (b) Surface energetics of polycrystalline $CuWO_4$ and WO_3 thin films derived from electrochemical and optical measurements.

Subsequent linear sweep voltammetry analysis was performed on polycrystalline $CuWO_4$ samples. As presented in Figure 4.a, a clear response to chopped AM1.5$_G$ illumination was observed, with a photocurrent density of approx. 400 μm.cm^{-2} at 1.6V vs. SCE and virtually no dark current up to this potential. At this stage of the research, $CuWO_4$ is clearly outperformed by reactively sputtered WO_3, known to generate photocurrent density of approx. 3.0 mA.cm^{-2} at 1.6V vs. SCE [2]. EIS analyses performed on $CuWO_4$- and WO_3-based PEC electrodes revealed that polycrystalline $CuWO_4$ owned a charge transfer resistance (R_{ct}=2,500 Ω.cm^2) one order of magnitude larger than that of pure WO_3 (R_{ct}=250 Ω.cm^2). In-depth understanding of $CuWO_4$ transport mechanisms will be required to reduce R_{ct} and achieve higher performing $CuWO_4$ PEC electrodes.

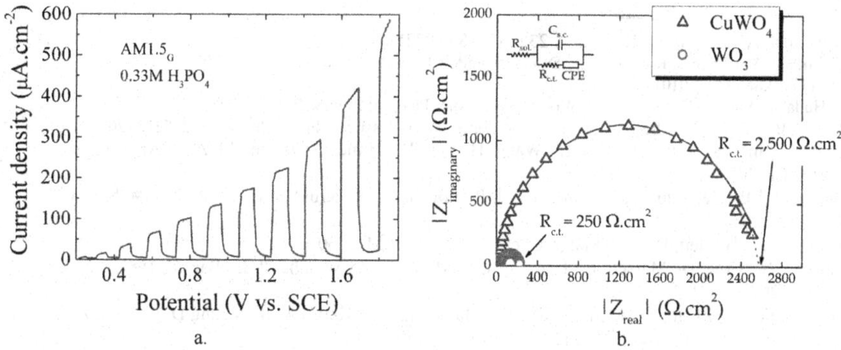

Fig. 4. (a) Linear sweep voltammetry analysis performed on polycrystalline $CuWO_4$ in 0.33M H_3PO_4 under chopped AM1.5$_G$ illumination. (b) Electrochemical impedance spectra measured on annealed $CuWO_4$ (blue triangles) and reactively sputtered WO_3 (red circles).

CONCLUSIONS

We reported in this communication on the photoelectrochemical properties of reactively co-sputtered $CuWO_4$. X-ray diffraction analysis revealed that a post-deposition annealing performed at 500°C in argon for 8 hours led to the crystallization of as-deposited (amorphous) thin film, a critical feature to obtain photo-activity. Electrochemical characterization pointed out that polycrystalline $CuWO_4$ thin films owned n-type conductivity, a V_{FB} of -0.35 V vs, SCE and a donor concentration of 1.20×10^{21} cm^{-3}. A side-by-side comparison of $CuWO_4$ and WO_3 band-edge positions indicated that copper tungstate was the metal oxide with the most favorable surface energetics for solar-assisted water splitting. Subsequent linear sweep voltammetry pointed out that $CuWO_4$ was indeed photoelectrochemically active to AM1.5_G illumination with a photocurrent density of approx. 400 $\mu m.cm^{-2}$ at 1.6V vs. SCE. Impedance spectra suggested that copper tungstate's main limitation could come from a rather large charge transfer resistance. Better understanding of its conduction mechanisms will improve $CuWO_4$ PEC performances.

ACKNOWLEDGMENTS

This work was supported by the U.S. Department of Energy under cooperative agreement DE-FC36-07GO17105. Financial support for A.B. by the Swiss National Science Foundation international project IZK0Z2-133944 "Oxide heterointerfaces in assemblies for photoelectrochemical applications" is gratefully acknowledged. The authors also would like to express their acknowledgments to Dr. Craig Jensen and Dr. Derek Birkmire for the valuable assistance in XRD measurements as well as Tina Carvalho from the University of Hawaii at Manoa for her help in SEM characterizations.

REFERENCES

[1] Fujishima, A., and *Honda*, K., *Nature*, 238, 37–38 (1972).
[2] N. Gaillard, Y. Chang, J. Kaneshiro, A. Deangelis and E. L. Miller, Proc. SPIE, Vol. 7770, 77700V; doi:10.1117/12.860970 (2010).
[3] M. Huda, Y. Yan, C.-Y. Moon, S.-H. Wei, M. Al-Jassim, Physical Review B 77, 1 (2008).
[4] B. Cole, B. Marsen, E. Miller, Y. Yan, B. To, K. Jones, M. Al-Jassim, J. Phys. Chem. C 112, 5213 (2008).
[5] Chen, L.; Shet, S.; Tang, H.; Ahn, K.; Wang, H.;Yan, Y.; Turner, J.; Jassim, M. A. J. Appl. Phys. 2010, 108, 043502 1-5.
[6] Doumerc, J.P.; Hejtmanek, J.; Chaminade, J. P.; Pouchard, M.; Krussanova, M. Phys. Stat. Sol. A 1984, 82, 285-294.
[7] Yourey J. E.; Bartlett, B. M. J. Mater. Chem. 2011, 21, 7651-7660.
[8] Gaillard, N.; Cole, B.; Marsen, B.; Kaneshiro, J.; Miller, E. L.; Weinhardt, L.; Bar, M.;Heske, C. J. Mater. Res. 2010, 25, 1-7.
[9] Zhu, J.; Wei, S.; Zhang, L.; Mao, Y.; Ryu, J.; Mavinakuli, P.; Karki, A. B.; Young D. P.; Guo, Z. J. Phys. Chem. C 2010, 114, 16335–16342.
[10] Huda, M. N.;Yan, Y.; Moon, C.; Wei S.; Jassim, M. M. Phys. Rev. B 2008, 77,195102-195114.
[11] Khaselev, O. K.; Turner, J. A., Science 1998, 280, 425-427.
[12] Khyzhun, O. Y.; Bekenev V. L.; Solonin, Y. M., J. Alloys Compd., 480 (2009), 184–189.
[13] M. V. Lalić, Z. S. Popović and F. R. Vukajlović, Comput. Mater. Sci., *2011*, 50, 1179.
[14] Arora, S. K.; Mathew T.; Batra, N. M. J. Phys. D: Appl. Phys. 1990, 23, 460.

Mater. Res. Soc. Symp. Proc. Vol. 1446 © 2012 Materials Research Society
DOI: 10.1557/opl.2012.1039

Oxygen Reduction Reaction Electrocatalytic Activity of SAD-Pt/GLAD-Cr Nanorods

Wisam J. Khudhayer [a*], Nancy Kariuki [b], Deborah J. Myers [b], Ali U. Shaikh [c], and Tansel Karabacak [d]

[a] Departments of Systems Engineering, [c] Chemistry, [d] Applied Science, University of Arkansas at Little Rock, Little Rock AR, 72204, USA

[b] Chemical Sciences and Engineering Division, Argonne National Laboratory, Argonne, IL 60439-4837, USA

ABSTRACT

Nanorod arrays of chromium (Cr) were grown on glassy carbon (GC) electrodes by a dc magnetron sputtering glancing angle deposition (GLAD) technique. The Cr nanorods were used as low-cost, high surface area, metallic supports for a conformal layer of Pt thin film catalyst, as a potential low-loading electrocatalyst for the oxygen reduction reaction (ORR) in polymer electrolyte membrane (PEM) fuel cells. A dc magnetron sputtering small angle deposition (SAD) technique was utilized for a conformal coating of Pt on Cr nanorods. The ORR activity of SAD-Pt/GLAD-Cr electrodes was investigated using cyclic voltammetry (CV) and rotating-disk electrode (RDE) techniques in a 0.1 M $HClO_4$ solution at room temperature. A reference sample consisting of GLAD Cr nanorods coated with a Pt thin film deposited at normal incidence ($\theta = 0°$) was prepared and compared with the SAD-Pt/GLAD-Cr nanorods. Compared to GLAD Cr nanorods coated with Pt thin film at $\theta = 0°$, the SAD-Pt/GLAD-Cr nanorod electrode exhibited higher ECSA and area-specific and mass-specific ORR activity. These results indicate that the growth of catalyst layer on the base-metal nanorods by the SAD technique provides a more conformal and possibly a nanostructured coating, significantly enhancing the catalyst utilization.

INTRODUCTION

The oxygen reduction reaction (ORR) at the cathode electrode of polymer electrolyte membrane (PEM) fuel cell is an important and well-studied electrochemical reaction because the slow kinetics of ORR causes it to be the dominant source of PEM voltage losses. [1] Typical cathode and anode electrocatalysts are platinum catalyst nanoparticles (3-5 nm in size) supported on carbon black. [2] In addition to the cost issue, this type of electrocatalyst faces other challenges related to the carbon support, summarized as follows: Oxidation of the carbon support causes catalyst loss, [3] the carbon support facilitates the formation of peroxide species that lead to degradation of the membrane polymer, [4] and carbon separates from the ionomer over time, leading to loss of catalyst utilization. [3] Therefore, extensive effort is currently underway to develop high-performance, durable, carbon-free, and low cost (low Pt loading) electrocatalyst materials. [1] For example, the 3M Company has demonstrated the improved durability and area-specific activity of nanostructured thin film Pt (NSTF Pt) electrocatalyst layers consisting of large-grained polycrystalline Pt thin film deposited on and encapsulating oriented crystalline whiskers of an organic pigment material.[5] A potential disadvantage of their support material, however, is the decomposition of the organic whiskers at elevated temperature (≥ 350 °C), which limits thermal treatment, such as catalyst annealing, to temperatures lower than 350 °C. [6] Bonakdarpour *et al.* [6] fabricated titanium (Ti) nanocolumns on glassy carbon (GC) disks using GLAD. The Ti nanocolumns were used as supports for Pt thin films deposited at a normal incidence angle ($\theta = 0°$). While the reported ORR specific activity (500 $\mu A/cm^2$ at 0.9 V) of these catalysts is lower than that reported for the 3M electrocatalyst (NSTF Pt), this activity is higher than that of the conventional Pt/C electrocatalyst. Furthermore, Gasda *et al.* deposited Pt thin films at normal incidence ($\theta = 0°$) on chromium nitride (CrN) nanoparticles and electrochemically-etchable carbon nanorod array supports fabricated by the GLAD technique. [7-9] Compared to the 3M NSTF Pt, one of the potential drawbacks of the above approaches [6-9] is that Pt tends to accumulate mainly on the tips of relatively dense nanoparticle/nanorod/nanocolumn supports, which results in lower catalyst utilization and dissolution of base metal, which in some cases can lead to complicated redox processes and deactivation of the electrode for ORR.

In this work, we investigate the use of a combination of GLAD and small angle deposition (SAD) techniques to fabricate chromium (Cr) nanorod array supports conformally coated with a thin layer of Pt. The GLAD method provides a novel capability for growing 3D nanostructure arrays with interesting material properties such as single crystallinity and the formation of uncommon crystal planes.[10-13] During GLAD, a flux of atoms produced by physical vapor deposition (e.g., sputtering) is incident on a rotating substrate (around its substrate normal) tilted at oblique angles, generally higher than 70°. This technique uses the "shadowing effect" which is a "physical self-assembly" through which some of the obliquely incident atoms may not reach certain points on the substrate due to the concurrent growth of parallel structures. [12,13] This leads to the formation of isolated nanostructures that can be in the shapes of rods, springs, zigzags, or spheres.[13] The conformal coating of base-metal GLAD nanorods with a thin layer of Pt is achieved using the SAD technique, developed by Karabacak et al.,[14,15] by controlling the substrate tilt angle. The main difference between SAD and GLAD techniques is that in SAD the incidence angle of the deposition flux is typically less than 45° and is adjusted based on the substrate surface pattern. In the SAD technique, the sidewalls and bottom portions of the support nanorods are effectively exposed to the deposition flux through the use of a small tilt angle and rotation of the substrate. This can lead to conformal coatings of very thin layers of catalyst on the support nanorods.[14]

In addition, since SAD uses a small substrate tilt angle, the incident flux of atoms will deposit on the sidewalls of the nanorods at high oblique angles as measured from the sidewall surface normal. For example, a nanostructured coating at the sidewalls can be achieved using a small substrate tilt angle, less than 30°, where the local incident angle on the sidewalls becomes larger than 70°, promoting the shadowing effect and formation of a nanostructured film. However, when the substrate tilt angle is too small and the deposition angle is close to normal incidence ($\theta \sim 0°$), most of the deposition occurs close to the tips of the nanorods, quickly clogging the gaps, which then leads to the formation of a continuous capping layer with poor coating of the sidewalls and bottoms of the nanorods.[16] Moreover, also due to the shadowing effect at the sidewalls during SAD, the film can have unusual crystallographic orientations. The crystal orientation can be controlled through the deposition parameters, such as tilt angle, substrate temperature, deposition rate, length and separation of nanorods, electrical biasing of the substrate, working gas pressure, and substrate rotation speed.

EXPERIMENTAL WORK

A DC magnetron GLAD sputtering system (Excel instruments, India) was employed for the fabrication of Cr nanorod arrays for a growth time of 15 minutes, which resulted in about 100 nm long Cr nanorods. The depositions were performed on glassy carbon disk inserts (5 mm OD x 4 mm thick, Pine Instrument) and silicon wafers using a 99.99 % pure Cr cathode (diameter ~ 2.54 cm). The substrates were mounted on the sample holder located at a distance of about 12 cm from the cathode. The base pressure of about 7.5 x 10^{-7} Torr was achieved using a turbo-molecular pump backed by a mechanical pump. The substrates were tilted so that the angle between the surface normal of the target and the surface normal of the substrate was 87°. The substrate was attached to a stepper motor and rotated at a speed of 2 rpm for growing vertical nanorods. During Cr deposition experiments, the power was 200 W with an ultra high purity Ar (99.99%) atmosphere at a pressure of 2.4×10^{-3} Torr. The deposition rate of GLAD Cr nanorods (i.e., nanorod length per growth time) was approximately 8.2 nm/min as measured by the analysis of cross-sectional SEM images.

After the deposition of the GLAD Cr nanorods, the substrates were tilted towards the Pt sputter source for coating with a Pt film by the SAD technique. Pt was deposited on the Cr nanorods for 60 seconds by utilizing dc sputter deposition from a 99.99% pure Pt cathode (diameter ~2.54 cm) at a deposition angle of $\theta = 30^\circ$. For comparison purposes, Pt was also deposited on GLAD Cr nanorods at normal incidence ($\theta = 0^\circ$) for 50 seconds. It should be noted that different deposition times were used to obtain similar Pt loadings on all the samples for comparison of their ORR activities. The substrates (arrays of Cr nanorods on GC and silicon wafer) were rotated around the substrate normal axis with a speed of 2 rpm. The deposition was performed under a base pressure of approximately 4×10^{-7} Torr. During Pt deposition experiments, the sputter power was 150 Watts with an Ar pressure of 3.2×10^{-3} Torr. The Pt loadings of the prepared samples were measured by quartz crystal microbalance (QCM, Inficon Q-pod QCM monitor, crystal: 6 MHz gold coated standard quartz). For the QCM measurements, Pt coatings were deposited directly on quartz crystals and the loading values were determined by comparing the oscillation frequencies of the blank and the coated crystal. [10]

The electrochemical tests (CV and RDE) were performed in deaerated and oxygen-saturated 0.1 M $HClO_4$ at room temperature for characterizing the ORR activity of the Pt/Cr nanorods electrocatalysts. The test setup was a typical three-electrode system (Pine Instrument bipotentiostat, North Carolina, USA), consisting of a working electrode (Pt/Cr nanorods on glassy carbon of 0.196 cm^2 area), a Pt wire counter electrode, and a Ag/AgCl reference electrode. The area of the working electrode samples was 0.196 cm^2, while the area of the counter electrode was higher to allow for a more uniform current density distribution through the working electrode. For CV measurements, the working electrodes were scanned from 0.05 to 1.20 V in O_2-free 0.1 M $HClO_4$ at a scan rate 50 mV/s. For chemical stability in the acidic environment, the electrodes were also scanned multiple times in the 0.6 to 1.2 V range at 50 mV/s. For the evaluation of ORR kinetics, RDE measurements were recorded between 0.05 and 1.05 V at a scan rate of 20 mV/s in O_2-saturated 0.1 M $HClO_4$ at room temperature and a rotation speed of 1600 rpm. To eliminate the effect of pseudo-capacitive currents on the calculated ORR activities, the background currents obtained using O_2-free 0.1 M $HClO_4$ were subtracted from the currents obtained in O_2-saturated electrolyte.

RESULTS AND DISCUSSION

Figure 1 shows the cross-sectional SEM images of the isolated vertical GLAD Cr nanorods before and after Pt deposition. Analysis of cross-sectional SEM images (Fig. 1a) showed the average length of Cr nanorods to be approximately 100 nm. A Pt thin film was deposited on GLAD Cr nanorods at $\theta = 0^\circ$ (normal incidence) and 30° (SAD deposition) as shown in Figs. 1b and c, respectively. The above samples will be referred to as Pt(0°)/Cr, Pt(30°)/Cr nanorods, respectively, in the remainder of the paper. Careful examination of Fig. 1b reveals that the Pt deposited at normal incidence is concentrated at the tips of the Cr nanorods, while a more uniform and conformal coating of Pt on the Cr nanorods was observed for the SAD-deposited Pt film, as shown in Fig. 1c. These observations were further confirmed by the EDX line mapping results, shown in Fig. 2, that were measured along the axis of the nanorods starting from their tips (i.e. tip being at the 0 nm position of Fig. 1). The Pt coverage percentage was estimated by the ratio of the coated portion, where the intensity of Pt is higher than that of Cr in Fig. 2, to the total length of nanorods. Figure 2a reveals that the Pt film deposited at normal incidence mainly accumulates at the tips of Cr nanorods with smaller amount deposited at the bottom. The estimated coverage of normal incidence deposited Pt on Cr nanorods is calculated to be ~30-40%. On the other hand, the coverage of SAD Pt is estimated to be about 75-80%. These results indicate that the SAD technique results in a more conformal thin film compared to normal incidence deposition.

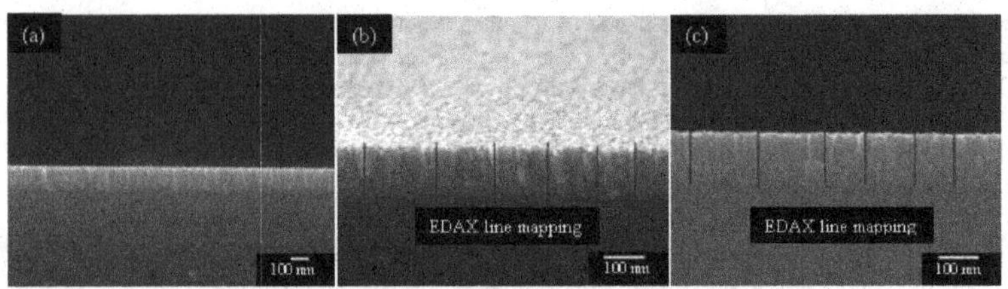

Figure 1: Cross-sectional scanning electron microscopy (SEM) images of (a) GLAD Cr nanorods, and Cr nanorods coated with a thin film of Pt deposited at (b) $\theta = 0^o$, and (c) 30° (SAD Pt).

Based on previously reported BET measurements on GLAD carbon nanorods, [17] Pt nanorods of 500 nm length are expected to have an effective surface area enhancement factor (SAEF, ratio of real surface area to substrate geometric area) of about 40. This can be used for estimating the SAEF for GLAD Cr nanorods coated SAD Pt thin films with the assumption that the total surface area of Cr/Pt nanorods is linearly dependent on their length. Based on that, the approximate SAEF of Pt/Cr nanorods was calculated to be ~8 (=40× (100/500)). Based on the estimated SAEF of Pt(30°)/Cr nanorods, the thickness of the Pt thin film on GLAD Cr nanorods is estimated to be approximately 3 to 5 nm, based on the calculated SAEF of Pt(30°)/Cr divided by the thickness of the thin film on a planar substrate.

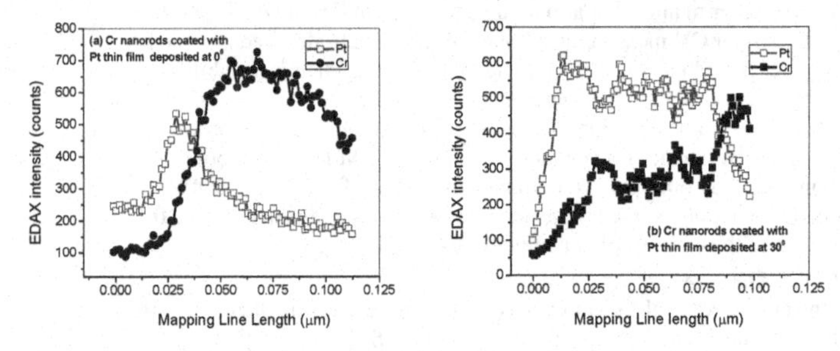

Figure 2: Average EDX elemental line mapping profiles of Pt and Cr along the GLAD Cr nanorods coated with a thin film of Pt deposited at (a) $\theta = 0^o$ and (b) 30° (SAD Pt), respectively. The line scanning starts from the top of the nanorods down to the nanorod-Si interface as shown in SEM images (Fig. 1).

Figure 3a compares the room temperature steady-state CV profiles of bare GLAD Cr nanorods and Cr nanorods coated with Pt thin film deposited at $\theta = 0^o$ and 30° with the same Pt loading of 0.04 mg/cm^2. The cyclic voltammogram of Cr nanorods shown in Fig. 3a agrees with those presented in the literature for Cr in acidic electrolyte. [18-21] The Pt thin films on Cr nanorods exhibit a shift in the onset potential for oxide formation compared to Pt alone (≥ 0.86 V vs. ~0.8 V for Pt alone, [11]), indicating an electronic interaction between the Pt and the Cr. The Pt(30°)/Cr electrode has an oxide reduction peak potential of 0.78 V, which is 50 mV more positive than the oxide reduction peak observed for Pt(0°)/Cr electrode, as shown in Fig. 3a. This can be attributed to a stronger electronic interaction between the thin, conformal Pt film with the Cr nanorods for the SAD thin film as compared to the thicker and less conformal films formed at 0°. [22] The greater conformality of the SAD film is also evidenced by the large CV peak intensities (i.e., a larger surface area is being exposed to the electrolyte) and relatively well-defined hydrogen adsorption/desorption peaks (enhanced electrochemical activity with different crystal planes). This strong interaction between the Pt and Cr may result in significant changes in the electronic states of surface Pt atoms inhibiting/reducing the formation of OH species which are considered poisoning species for ORR. [23-26]

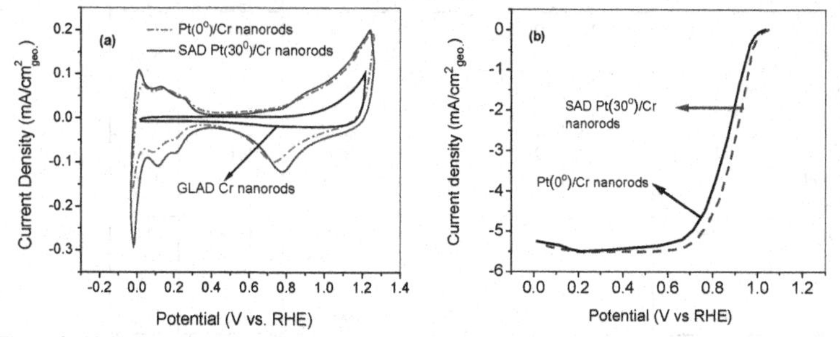

Figure 3: (a) Room temperature CVs in O_2-free 0.1 M $HClO_4$ at a scan rate 50 mV/s for GLAD Cr nanorods coated with a thin film of Pt deposited at $\theta = 0°$ and 30° (SAD Pt) all with a Pt loading of 0.04 mg/cm^2. The solid black line shows the CV profile of GLAD Cr nanorods without the Pt coating and (b) Room temperature RDE profiles in oxygen-saturated 0.1 M $HClO_4$ at a scan rate 20 mV/s and rotation speed 1600 rpm for GLAD Cr nanorods coated with a thin film of Pt deposited at $\theta = 0°$ and 30° (SAD Pt).

The ECSA of the prepared electrodes was determined by integrating the charge in the hydrogen adsorption region derived from CV profiles between the double layer region and the onset of hydrogen evolution, after subtracting the double layer charge. The Pt(30°)/Cr electrode exhibits an ECSA of 16 m^2/g$_{Pt}$ compared to 12.2 m^2/g$_{Pt}$ for the films deposited at 0°. The larger ECSA for the SAD film as compared to the films deposited at 0° further supports the SEM and EDX results discussed above. The RDE results in Fig. 3b show that the Pt(30°)/Cr nanorods exhibit an $E_{1/2}$ of ~0.9 V, while the GLAD Cr nanorods coated with Pt thin films at $\theta = 0°$ has an $E_{1/2}$ of ~0.86 V. The ORR kinetic currents (I_k) at 0.9 V were extracted from these RDE data using the well-known mass-transport correction for RDE measurements. [2] It should be noted that the assumptions implicit in this equation valid over the current range of $0.1\ I_{lim} < I < 0.8\ I_{lim}$ and thus this equation is valid for calculating I_k at 0.9V. [25] The area- (*SA*) and mass- (*MA*) specific activities are derived by normalizing the kinetic currents by the ECSA and the Pt loading of the catalyst applied to the electrode, respectively. At 0.9 V, the *SA* of GLAD Cr nanorods coated with the SAD Pt thin film deposited at $\theta = 30°$ is approximately 885 μA/cm^2, which is about 68% higher than that the GLAD Cr nanorods coated with the Pt thin film deposited at 0° with *SA* of 525 μA/cm^2, as shown in Table I. It can be seen from Table I that the GLAD Cr nanorods coated with the SAD Pt thin film have ORR mass activities normalized to Pt loading of ~0.150 A/mg, which is higher than the ~0.064 A/mg observed for Pt(0°)/Cr. The improved activity might be attributed to a better Pt conformality, especially at the sidewalls of the nanorods, and a preferential exposure of certain crystal facets. [11]

Table I: Summary of the evaluated electrocatalytic activity (area-specific and mass-specific activities at 0.90 V) of Cr nanorods coated with SAD ($\theta = 30°$) Pt thin film catalyst in 0.1 M $HClO_4$ and comparison to the literature values for various Pt-based catalysts.

Method	Sample	Pt loading (mg/cm^2)	ECSA (m^2/g)	T (°C)	Scan rate (mV/s)	I_s ($\mu A/cm^2$)	I_m (A/mg)	Ref.
RDE	Pt(0°)/Cr nanorods	0.040	12.2	20	20	525	0.064	This work
RDE	Pt(0°)/Ti nanorods	50 nm thick Pt film	SEF* ~ 11	21	5	500	-	[6]
RDE	SAD Pt(30°)/Cr nanorods	0.040	16	20	20	885	0.150	This work
RDE	200 nm long Pt nanorods	0.160	12	20	10	1,080	0.130	[11]
TF-RDE	20% Pt/C	0.020	60	20	20	288	0.180	[11]
TF-RDE	NSTF Pt	0.042	8	20	20	750	0.060	[25]

*SEF: Surface Enhancement Factor [6]

The ORR specific activity of Pt(30°)/Cr is compared with the literature values for conventional Pt/C, solid GLAD Pt nanorods, and nanostructured thin film Pt (3M NSTF Pt) electrocatalysts (Table I). It was found that the Pt(30°)/Cr nanorod array electrode exhibits higher *SA* than conventional Pt/C and 3M Pt NSTF electrocatalysts. This enhancement might be attributed to (1) the dominance of the preferred crystal orientation of the SAD Pt coating at the nanorod sidewalls for ORR [11] and (2) the change in the electronic states of surface Pt atoms induced by the electronic interaction between Pt and Cr nanorods which reduces/inhibits the formation of OH species which are considered poisoning species for ORR. [23-26] In addition to the above possible advantages, the higher *SA* of Pt(30°)/Cr electrode compared to Pt/C and 3M Pt NSTF catalysts might be arising from its larger crystallite size, in the range of 5-100 nm (i.e., film thickness–planar size along the Cr nanorod length), which results in average coordination numbers of surface Pt and corresponding surface electronic properties approaching those of polycrystalline Pt. [11,25] Furthermore, the mass-activity (*MA*) of Pt(30°)/Cr nanorod array electrode is higher than that of the 3M Pt NSTF, based on the reported area-specific activity at 0.9 V and ECSA of 8 m^2/g, but it is still lower than the reported values in the literature [2,25] for Pt/C electrocatalyst. [11] Compared to the solid GLAD Pt nanorods with *MA* of 0.13 A/mg, [11] A significant enhancement in *MA* value for the Pt(30°)/Cr electrode is expected since a small amount of Pt was used to coat the base nanorod support (GLAD Cr nanorods). The small *MA* enhancement of 0.02 A/mg is attributed to the incomplete coating of GLAD Cr nanorods by SAD Pt thin film, especially at the bottom of the nanorods as shown by EDX analysis above, and to the excessive thickness of the Pt film. In addition, there is a possibility that the crystallographic structure of the sidewalls of Pt(30°)/Cr nanorods is different that of solid GLAD Pt nanorods where the ORR activity is dominated by Pt(110), which is the most active surface for ORR. [11] We believe that the *MA* parameter can be further improved by increasing the separation of Cr nanorods so that the SAD approach can be more effective in reaching the bottom of the nanorods, resulting in a more conformal coating of Pt thin film on the nanorods and a better Pt utilization. This approach may also enhance the electrode porosity resulting in better oxygen transport within the catalyst layer.

CONCLUSION

SAD-Pt/GLAD-Cr nanorods exhibit higher electrochemically-active surface area (ECSA) and area- and mass-specific ORR activities as compared to GLAD Cr nanorods coated with a Pt thin film deposited at normal incidence ($\theta = 0°$). The improved ORR activity and enhanced catalyst utilization of SAD-Pt/GLAD-Cr electrode might be attributed to a better Pt conformality, especially at the sidewalls of the nanorods, and a preferential exposure of certain crystal facets.

ACKNOWLEDGEMENTS

The authors would like to thank the UALR Nanotechnology Center and Dr. Fumiya Watanabe for his valuable support and discussions during SEM/EDX and XRD measurements. The Argonne National Laboratory authors would like to thank the Department of Energy, Office of Energy Efficiency and Renewable Energy, Fuel Cell Technologies Program (Nancy Garland, DOE Program Manager). Argonne is a U.S. Department of Energy Office of Science Laboratory operated under Contract No. DE-AC02-06CH11357 by U Chicago Argonne, LLC.

REFERENCES

1. H. A. Gasteiger, S. S. Kocha, B. Sompalli, F. T. Wagner, *Applied Catalysis B: Environmental* **56**, 9-35 (2005).
2. J. Zhang, *"PEM fuel cell electrocatalysts and catalyst layer: fundamentals and applications"*, Springer-Verlag London Limited, 2008.
3. H. Tang, Z. G. Qi, M. Ramani, J. F. Elter, *J. Power Sources* **158**, 1306 (2006).
4. J. L. Qiao, M. Saito, K. Hayamizu, T. Okada, *J. Electrochem. Soc.* **153**, A967 (2006).
5. M. K. Debe, A. K. Schmoeckel, S. M. Hendricks, G. D. Vernstrom, G. M. Haugen and R. T. Atanasoski, *ECS Transactions* **1** (8), 51-66 (2006).
6. A. Bonakdarpour, M. D. Fleischauer, M. J. Brett, J. R. Dhan, *Applied Catalysis A: General* **349**, 110-115 (2008).
7. M. D. Gasda, G. A. Eisman, D. Gall, *Journal of Electrochemical Society* **157** (1), B71-B76 (2010).
8. M. D. Gasda, G. A. Eisman, D. Gall, *Journal of Electrochemical Society* **157** (1), B113-B117 (2010).
9. M. D. Gasda, G. A. Eisman, D. Gall, *Journal of Electrochemical Society* **157** (3), B437-B440 (2010).
10. W. J. Khudhayer, Ali U. Shaikh, and Tansel Karabacak, *Advanced Science Letters* **4**, 3551-3559, 2011.
11. W. J. Khudhayer, Nancy Kariuki, Xiaoping Wang, Deborah J. Myers, Ali U. Shaikh, and Tansel Karabacak, *Journal of Electrochemical Society* **158** (8), B1029-B1041 (2011).
12. T. Karabacak and T.-M. Lu, in *Handbook of Theoretical and Computational Nanotechnology*, edited by M. Rieth and W. Schommers (American. Scientific Publishers, Stevenson Ranch, CA, 2005), chap. 69, p. 729.
13. T. Karabacak, G.-C. Wang, and T.-M. Lu, *J. Vac. Sci. Technol.* **A 22**, 1778 (2004).
14. T. Karabacak and T.-M. Lu, *J. Appl. Phys.* **97**, 124504 (2005).
15. T. Karabacak and T.-M. Lu, *U.S. Patent No*: 7,244,670, Jul 17 (2007).
16. A. Dolatshahi-Pirouz, T. Jensen, T. Vorup-Jensen, Rikke Bech, J. Chevallier, F. Besenbacher, M. Foss, and D.S. Sutherland, *Adv. Eng. Mater.* **12**, 899 (2010).
17. G. K. Klema and M. J. Brett, J. of electrochemical Society **150**, p. E342 (2003).
18. T. P. Moffat, R. M. Latanision, *J Electrochem. Soc.* **139**, 1869–1879 (1992).
19. M. Bojinov, G. Fabricius, T. Laitinen, T. Saario, G. Sundholm, *Electrochim. Acta* **44**, 247–261 (1998).
20. V. Maurice, W. P. Yang, P. Marcus, *J Electrochem. Soc.* **141**, 3016–3027 (1994).
21. J. H. Gerretsen, J. H. W. De Wit, *Corrosion Sci.* **30**, 1075–1084 (1990).
22. S. Mukerjee, and S. Srinivasan, J. Electroanal. Chem. 357, 201-224 (1993).
23. V. Stamenkovic, T. J. Schmidt, P. N. Ross, and N. M. Markovic, J. *Phys. Chem.* **B 106**, 11970-11979 (2002).
24. V. V. Stamenkovic, T. J. Schmidt, P. N. Ross, and N. M. Markovic, J. *of Electroanalytical Chemistry* **554-555**, 191-199 (2003).
25. K. J. J. Mayrhofer, B. B. Blizanac, M. Arenz, V. R. Stamenkovic, P. N. Ross, and N. M. Markovic, *J. Phys. Chem.* **B 109**, 14433-14440 (2005).
26. V. R. Stamenkovic, B. S. Mun, M. Arenz, K. J. J. Mayrhofer, C. A. Lucas, G. Wang, P. N. Ross, and N. M. Markovic, *Nature Materials* **6**, 241-247 (2007).

Mater. Res. Soc. Symp. Proc. Vol. 1446 © 2012 Materials Research Society
DOI: 10.1557/opl.2012.916

STUDY OF CO-ASSEMBLED CONDUCTING POLYMERS FOR ENHANCED ETHANOL ELECTRO-OXIDATION REACTION

Le Q. Hoa,[1] Hiroyuki Yoshikawa,[1] Masato Saito,[1] and Eiichi Tamiya[1]
[1]Department of Applied Physics, Graduate School of Engineering, Osaka University,
2-1 Yamadaoka, Suita, Osaka 565-0871, Japan

ABSTRACT

Herein, we investigated the effects of polyaniline (PANI) and polypyrrole (PPY) in their native and co-assembled forms as a thin layer on Pt nanoparticle-decorated multi-walled carbon nanotubes (Pt/MWCNTs) toward the ethanol oxidation reaction (EOR). The co-assembled conducting PANI-PPY deposited Pt/MWCNTs was successfully synthesized and demonstrated significant enhancement of the electro-catalytic activity and stability toward EOR as revealed by electrochemical characterizations. The presented results indicate that in the co-assembled form, PANI and PPY retained their own superior effects on the enhancement of stability and catalytic activity via intermediate species removal and ethanol adsorption, respectively. This preliminary result reveals a new strategy for the use of conducting polymers as potential catalyst supports due to its facile fabrication and functionalization, cost effectiveness and environmental friendliness in comparison to alloys and metal oxides, factors which are necessary for the practical application of direct ethanol fuel cells in the near future.

INTRODUCTION

Prospecting direct ethanol fuel cells as practical bio-energy technology is limited by the slow kinetics of the inefficient ethanol oxidation reaction (EOR), low durability of catalysts due to contamination by intermediate species, and last but not least, difficulties in fuel cell fabrication procedures and cost effectiveness for large-scale implementation [1, 2].

To date, researchers have tried to explore the use of non-noble metals as catalysts for the ethanol oxidation process [3-10], and platinum (Pt) remains an irreplaceable component due to its incomparably high activity, especially in neutral and acidic condition [2]. The problem is that Pt-based catalysts are readily poisoned by the -CO_{ads} or -CH_x intermediates generated from the incomplete oxidation of ethanol [5, 14]. To deal with such matters, many authors employed the use of Pt-based alloys containing tin (Sn), tin oxides, or other transition metals and archived the improvement on both the catalytic capacity and the stability of catalysts [5-7]. These enhancements are derived firstly from Sn or SnO_x, which provides a surface rich in oxygen containing species capable of removing adsorbed CO by facilitating its oxidation to CO_2. Secondly, the "OH-carpet" created by SnO_x is able to preclude Pt-M (e.g. Rh, Ru) from reacting with water to form M-OH, making them in low-coordination states and available for ethanol oxidation [7]. In all of the above mentioned cases, the cost effectiveness and difficulty of large-scale fabrication are largely ignored.

Taking into understanding these alloy effects, our approach makes use of the organic species as a support for conventional Pt nanoparticles, rather than metal oxides and alloys. In this method, it is imperative that the organic species is not only electroactive but also conductive, hence conducting polymers are promising candidates. Recently, Pandey et al. have used PANI and poly (3,4-ethylenedioxythiophene) (PEDOT) to enhance the electrocatalytic activity of Pd on a gold substrate for EOR in alkaline conditions. Their work demonstrated the importance of conducting polymers in forming an ideal conducting matrix [8, 9]. However, these improvements were not credited to the conducting polymers themselves, but the -OH groups attached on gold substrate [9]. Herein, we particularly investigated the effect of PANI and PPY in their native and co-assembled forms not as a matrix but as a top-layer of Pt/MWCNTs on EOR. Interestingly, the data showed that the co-assembled polymers could significantly enhance the electrocatalytic activities and stability of Pt/MWCNTs despite the lower activities observed in each individual component.

EXPERIMENT DETAILS

Reagent grade aniline, concentrated nitric acid, glycerol, camphorsulfonic acid (CSA), ammonium peroxydisulfate (APS) and Multi-walled carbon nanotubes (MWCNTs, purity $\geq$ 95%, 40-70 nm in diameter) were obtained from Wako Co., Japan. The electrocatalyst precursor ($H_2PtCl_6 \bullet 6H_2O$) and poly(pyrrole) in doped state (5% in water) were purchased from Aldrich, USA. MWCNTs were surface oxidized by refluxing in conc. HNO_3 solution at 140°C-160°C for 24 h, and then decorated with Pt-NPs by refluxing with $H_2PtCl_6 \bullet 6H_2O$ in aqueous glycerol at 140°C, pH 10 for 12 hours followed by filtering and vacuum drying [19-22]. The chemical polymerization of PANI was carried out at room temperature using a rapid mixing method [20, 22]. Fabrication of the catalyst on C paper substrates was done by dispersing 5 mg Pt/MWCNTs into 1 mL ethanol and 50 µl nafion (5%, Wako Co., Japan), which was then casted onto the surface of carbon paper to obtain a total Pt loading of 50 µg cm^{-2}. The electrodes were dried and used for conducting polymer deposition by direct casting. The concentration of dispersed PANI and PPY was 5 mg mL^{-1} of ethanol 99.5% and deionized water, respectively. The co-assembled PANI-PPY suspension was made by simply mixing these two native polymer solutions with varying volume ratios to get different weight ratios between PANI and PPY. The electrodes were then subjected to thermal treatment at 80°C for 1 hour. In all cases, Pt loading was 0.05 mg cm^{-2}. The surface morphologies of various conducting polymer-supported Pt-MWCNTs on C paper were acquired by scanning electron microscopy (SEM) (DB 235 microscope, FEI). Electrochemical measurements were performed with an Autolab potentiostat/galvanostat PGSTAT12 (EcoChemie, Netherlands).

DISCUSSION

Physical and morphological characterization of PANI-PPY on Pt/MWCNTs/C

In this study, "native form" of PANI and PPY is used to denote the polymer itself as a sole component and "co-assembled form" PANI-PPY is used to denote the mixture of PANI and PPY with varied weight ratios.

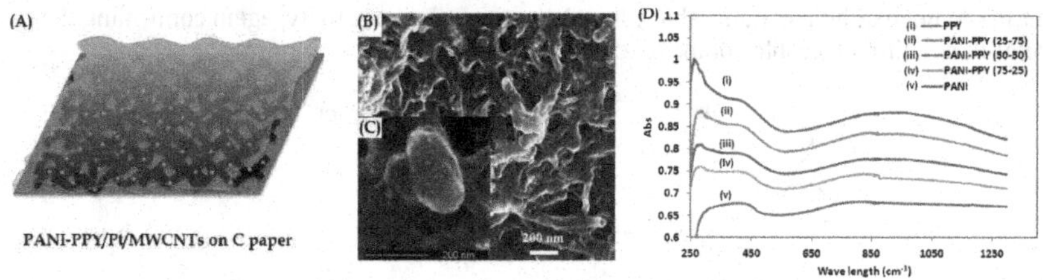

Figure 1. Illustration (A) and SEM images (B and C) of PANI-PPY on Pt/MWCNTs modified carbon paper. (D) UV–vis absorption spectra of the aqueous dispersions of various weight ratio PANI-PPY compared to pure PANI and pure PPY in HCl 1M.

The Pt/MWCNTs layer (0.25 mg cm^{-2} containing 0.05 mg Pt cm^{-2}) was first casted onto carbon paper resulting in a porous platform with high surface area. Next, a conducting polymer, either in native form or in co-assembled form, was deposited on Pt/MWCNTs layer and then a thermal treatment was applied to strengthen the binding between Pt/MWCNTs and the conducting polymers. After heating, SEM images (figure 1B and C) evidently suggest that the conducting polymer was melted and coated on Pt/MWCNTs. This layer was sufficiently thin that Pt nanoparticles could be observed as white dots impregnated on the carbon nanotube-like structure (figure 1C). Concerning this aspect, the loading of the polymer needed to be well-controlled such that the Pt active surface was not buried and inhibited by the polymer layer. In native form, less than 125.0 µg PANI cm^{-2} and 62.5 µg PPY cm^{-2} loading resulted in partly covered Pt/MWCNTs/C and therefore unstable electrochemical data, thus it is not presented here. Increasing the polymer loading caused a significantly thick layer covered Pt/MWCNTs, thus inhibiting the Pt active sites, resulting in lower current density of ethanol oxidation peaks. Consequently, 125 µg cm^{-2} is the optimized loading of PANI and PPY. UV–vis absorption was used to analyze the electronic characteristics of native PANI, PPY, and various weight-ratio mixturing PANI-PPY solutions. Samples were prepared in a 1 M HCl solution because a low pH was required to maintain the protonated forms of PANI and PPY, which were necessary for conductivity [17, 18]. As shown in figure 1D, all samples exhibited two characteristic absorption bands at approximately 430 nm and 800 nm, corresponding to a polaron–π^* transition and a π–polaron transition, respectively. Since the mixture of PANI and PPY does not show any new peaks, the spectrum of the mixture could be considered as the physical combination and average of the two individual spectra. Therefore, we can conclude that the action of mixing PANI and PPY did not cause any modification to the chemical structure or the electronic states of either polymer. Since the physical properties of the mixture are similar to those of the invidual polymers, it is expected that when deposited onto the Pt/MWCNTs surface, the co-assembled conducting polymer PANI-PPY would possess the characteristics of both polymers, which could therefore provide a synergistic catalytic effect on the enhancement of ethanol oxidation reaction of the synthesized electrode.

Catalytic performance

The cyclic voltammograms (CVs) of conducting polymers in their native and co-assembled forms deposited on Pt/MWCNTs/C in comparison with Pt/MWCNTs/C and conducting polymers on bare carbon paper as control samples are shown in figure 2. All control samples (without Pt) do not exhibit any noticeable ethanol electrooxidation activity, again confirming the key role of Pt as an irreplaceable component.

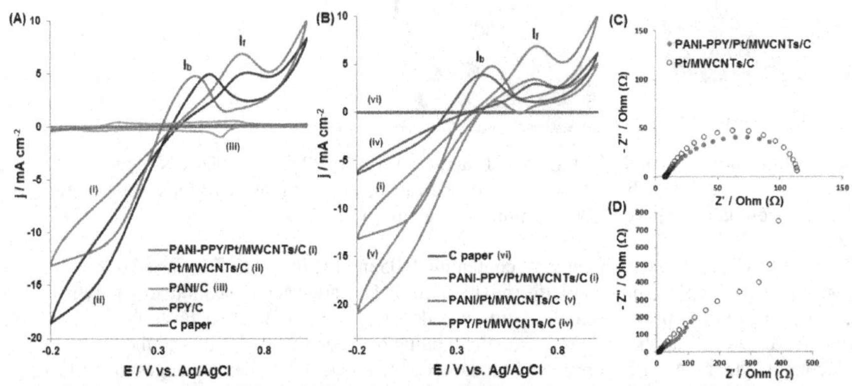

Figure 2. The electro-catalytic activities of co-assembled conducting polymers deposited Pt/MWCNTs/C in comparison with native polymers deposited Pt/MWCNTs/C and control samples including Pt/MWCNTs/C, PANI/C, PPY/C and C paper. (A) and (B) are the 30th cyclic voltammograms in 0.5 M C_2H_5OH and 0.1 M $HClO_4$ with scan rate 50 mV s^{-1}. Electrochemical impedance spectra at (C) 0.4 V and (D) 0.6 V applied potentials.

As shown in figure 2A, the co-assembled PANI-PPY/Pt/MWCNTs/C with the same Pt loading exhibits a maximum current density of approximately 1.5 times higher than that of Pt/MWCNTs/C, and the oxidation peak is ~25 mV negatively shifted. Furthermore, the reaction kinetics in the first range of potential (less than 0.5 V) and therefore the onset potential in case of PANI-PPY/Pt/MWCNTs/C are also much more rapid than Pt/MWCNTs/C. Electrochemical impedance spectra within this potential range (figure 2C) also verify the higher charge-transfer resistance, defined by larger semicircles at high frequency, in case of Pt/MWCNTs/C than that of PANI-PPY/Pt/MWCNTs/C. This means that the co-assembled conducting polymer does promote the ethanol oxidation reaction on Pt active sites at an earlier voltage and a faster charge-transfer rate during the oxidation reaction than Pt by itself. For a deeper understanding of the supporting roles of each polymer over whole reaction on the Pt surface, a closer examination of native polymers coated Pt/MWCNTs/C gave us some important clues (figure 2B). Firstly, in the forward scan of PPY/Pt/MWCNTs/C, the current density in the onset potential range (-0.2 V to 0.5 V) when ethanol starts adsorbing onto the Pt surface and dissociating is significantly higher than that of both Pt/MWCNTs/C and PANI-PPY/Pt/MWCNTs/C, indicating that the presence of PPY contributed to the first part of the reaction. It is possible that the hydrophobic PPY layer adsorbs ethanol much more readily than water [19], thereby enhancing the concentration of ethanol on adjacent Pt active sites. Secondly, the current density ratio between the ethanol oxidation peak in the forward scan (I_f) and the poisoning intermediate oxidation peak in the backward scan (I_b) of PANI/Pt/MWCNTs/C is much higher than any other case, highlighting PANI's effect on reducing catalyst poisoning from the adsorbents. In contrast to PPY, PANI is more hydrophilic and could adsorb water on positive $+NH^{-}$ sites, thus it may enhance the oxidation of CO_{ads} or other intermediates via the bi-functional mechanism much like Sn or SnO_x

[14]. It is noteworthy that the I_b in the case of PPY/Pt/MWCNTs is dramatically increased compared to I_f, and the current density in the onset potential range of PANI/Pt/MWCNTs is lower than that of Pt/MWCNTs/C. Thus, the advantage of PANI is the disadvantage of PPY and vice versa. Consequently, the co-assembled PANI-PPY deposited Pt/MWCNTs/C demonstrated the average magnitude of both advantages and disadvantages of PANI and PPY on the kinetic reaction and I_f/I_b ratio, where the current density was remarkably increased. In the impedance analysis of Pt/MWCNTs/C at 0.6 V, the Nyquist plots start flipping toward the second quadrant of the complex plane, indicating the passivation of the electrode surface, while PANI-PPY/Pt/MWCNTs/C remained at observably higher knee frequencies (figure 2D). It further confirmed that the co-assembled conducting polymer does contribute to the depoisoning of the catalyst surface. The remarkably increased current density of ethanol oxidation peak in the case of PANI-PPY/Pt/MWCNTs/C revealed that the enhanced catalytic activity rests on the resonant synergy between these two conducting polymers and their interaction with the Pt surface. These results also emphasized the importance of controlling the ratio between PANI and PPY on the outcome of electro-catalytic activities.

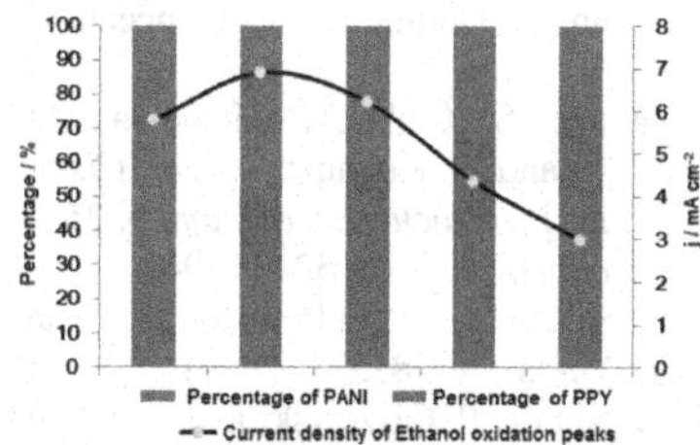

Figure 3. Effect of weight ratio between PANI and PPY in the co-assembled form (deposited on Pt/MWCNTs/C) on maximum current density of ethanol oxidation with the measurement condition is the same as in figure 2A.

Table 1. Contribution of each polymer to the total performance of the co-assembled PANI-PPY electrode

	Ethanol adsorption	De-poisoning capability	Total performance
PANI	-	+++	++ (2nd)
PPY	+++	-	+ (3rd)
PANI-PPY	++	++	+++ (1st)

As evidently shown in figure 3, the optimum weight ratio for PANI-PPY to gain the highest current density is 1:1. Increasing either PANI or PPY partials resulted in the decrease of catalytic activity. These results are in agreement with the data and suggestion in the previous section in that two invidual components combine and exhibit their average characteristics in the co-assembled form.

CONCLUSIONS

In summary, the co-assembled conducting PANI-PPY deposited Pt/MWCNTs was successfully synthesized and demonstrated significant enhancement of the electro-catalytic activity and stability toward the ethanol oxidation reaction. The current density was increased by a factor of 1.5 and both the onset potential and oxidation peak potential were negatively shifted. The presented results indicate that in the co-assembled form, PANI and PPY retained their own superior effects on the enhancement of stability and catalytic activity via intermediate species removal and ethanol adsorption, respectively. Detailed studies on the mechanistic aspects and further structural and functional optimizations are now in progress.

ACKNOWLEDGMENTS

This work was supported by Japan Science & Technology (JST), CREST.

REFERENCES

1. C. Lamy, S. Rousseau, E. M. Belgsir, C. Coutanceau and J. -M. Léger, *Electrochim. Acta* **49**, 3906 (2004).
2. C. Lamy, A. Lima, V. L. Rhun, F. Delime, C. Coutanceau and J. -M. Léger, *J. Power Sources* **105**, 285 (2002).
3. L. Yang, S. Kinoshita, T. Yamada, S. Kanka, H. Kitagawa, M. Tokunaga, T. Ishimoto, T. Ogura, R. Nagumo, A. Miyamoto and M. Koyama, *Angew. Chem. Int. Ed.* **49**, 5348 (2010).
4. L. Cao, G. Sun, H. Li and Q. Xin, *Electrochem. Commun.* **9**, 2541 (2007).
5. T. Iwasita and E. Pastor, *Electrochim. Acta* **13**, 534 (1994).
6. A. Kowal, M. Li, M. Shao, K. Sasaki, M. B. Vukmirovic, J. Zhang, N.S. Marinkovic, P. Liu, A. I. Frenkel and R. R. Adzic, *Nat. Mater.* **8**, 328 (2009).
7. A. Kowal, S. Lj. Gojkovic, K. S. Lee, P. Olszewski and Y. E. Sung, *Electrochem. Commun.* **11**, 724 (2009).
8. R. K. Pandey and V. Lakshminarayanan, *J. Phys. Chem. C* **113**, 21596 (2009).
9. R. K. Pandey and V. Lakshminarayanan, *J. Phys. Chem. C* **114**, 8511 (2010).
10. S. Biallozor, A. Kupniewska and V. Jasulaitene, *Fuel Cells* **3**, 8 (2003).
11. X. -H. Jian, D. -S. Tsai, W. -H. Chung, Y. -S. Huang and F. -J. Liu, *J. Mater. Chem.* **19**, 1606 (2009).
12. S. J. Hwang, S. J. Yoo. T. -Y. Jeon, K. -S. Lee, T. -H. Lim, Y. -E. Sung and S. -K. Kim, *Chem. Comm.* **46**, 8401 (2010).
13. M. Watanabe and S. Moto, *J. Electroanal. Chem.* **60**, 267 (1975).
14. H. Okuzaki, T. Kondo and T. Kunugi, *Polymer* **40**, 997 (1999).
15. W. Cheung, P. L. Chiu, R. R. Parajuli, Y. Ma, S. R. Ali and Huixin He, *J. Mater. Chem.* **19**, 6475 (2009).
16. B. Weng, R. L. Shepherd, K. Crowley, A. J. Killard and G. G. Wallace, *Analyst* **135**, 2783 (2010).
17. P. Jiménez, P. Castell, R. Sainz, A. Ansón, M.T. Martínez, A.M. Benito, and W.K. Maser, *J. Phys. Chem. B* **114**, 1581 (2010).
18. M. Zhao, X. Wu, and C. Cai, *J. Phys. Chem. C* **113**, 4989-4991 (2009).
19. L. Q. Hoa, Y. Sugano, H. Yoshikawa, M. Saito, and E. Tamiya, *Biosens. Bioelectron.* **25**, 2509 (2010).
20. L. Q. Hoa, H. Yoshikawa, M. Saito, and E. Tamiya, *J. Mater. Chem.* **21**, 4068 (2011).
21. L. Q. Hoa, M. C. Vestergaard, H. Yoshikawa, M. Saito, and E. Tamiya, *Electrochem. Commun.* **13**, 746 (2011).
22. L. Q. Hoa, Y. Sugano, H. Yoshikawa, M. Saito, and E. Tamiya, *Electrochim. Acta* **56**, 9875 (2011).

Mater. Res. Soc. Symp. Proc. Vol. 1446 © 2012 Materials Research Society
DOI: 10.1557/opl.2012.953

Preparation and Characterization of Platinum/Ceria-based Catalysts for Methanol Electro-oxidation in Alkaline Medium

Christian L. Menéndez[1] Ana-Rita Mayol[2] and Carlos R. Cabrera[1,3]

[1]University of Puerto Rico, Rio Piedras Campus, Chemistry Graduate Program, PO Box 23346, San Juan, PR, 00931-3346, U.S.A.
[2]Institute for Functional Nanomaterials, University of Puerto Rico, Rio Piedras Campus, PO Box 23346, San Juan, PR 00931-3346, U.S.A.
[3]Center for Advanced Nanoscale Materials, University of Puerto Rico, Rio Piedras Campus, PO Box 23346, San Juan, PR 00931-3346,U.S.A.

ABSTRACT

Platinum/Cerium oxide-based catalysts for methanol electro oxidation were prepared by the occlusion deposition technique. Composite glassy carbon (GC) electrodes were modified and then tested towards the methanol electro oxidation half reaction in acid and alkaline medium. Cyclic voltammetry and chronoamperometry techniques were used to test the catalytic response of the composite electrodes. AFM studies were carried out in order to have a measurement of the particle size and distribution of the platinum/ceria catalyst on HOPG.

INTRODUCTION

The conversion of the chemical energy of alcohols into electricity has been carried out using two basic types of fuel cells: alkaline and acidic. The commercialization of acidic-type fuel cells still remains as a problem due to the large amount of platinum required to obtain good performance. The use of alkaline-type fuel cells is more promising because the catalytic activity to the methanol oxidation reaction is higher[1].

In most of the investigations about fuel cells methanol is used as a fuel. The principal reason of this is because of the big energy density associated with this alcohol (6 $kWhKg^{-1}$), the low cost of the alcohol and the easy storage and transportation in comparison with other fuels as hydrogen gas. In addition, methanol is the shortest alcohol and its oxidation is more feasible than for other alcohols since methanol does not have C-C bonds [2].

The use of alkaline medium in fuel cells technology has increased due to its higher stability and efficiency, versatility of materials and catalysts for the electrodes and more efficient reactions at the cathode and the anode. The electro oxidation of methanol in alkaline medium has made possible the use of nickel as catalysts [3].

To this day, Pt-Ru is the most efficient and stable catalyst for methanol electro oxidation at low temperature fuel cells. This is the reason why different bimetallic combinations of catalysts with

non noble elements have been studied and tested in order to enhance the catalytic activity of methanol electro oxidation [4].
Cerium oxide is one of the most used rare earth oxide catalysts due to its good efficiency to oxidize carbon monoxide. Pt-CeO_2/CNTs (carbon nanotubes) catalysts have been prepared and compared their efficiency towards the methanol electro oxidation with Pt only supported on CNTs. Results obtained indicate that CeO_2 can significantly increase the catalytic activity of Pt for this reaction [5].
We have reported previously that the use of cerium oxide as catalyst support for acid-type methanol oxidation enhances the catalytic activity when using a GC/Pt/CeO_2 as catalyst in comparison with a GC/Pt only composite electrode [6]. The GC/Pt/CeO_2 electrodes were prepared using the occlusion electro deposition method. In a similar way, Díaz [7] and co workers used ceria to prepare ceria-Pt composite electrodes for acidic-type alcohol electro-oxidation and showed that ceria appears to be an excellent alternative to replace ruthenium in direct alcohol fuel cells. In this paper we propose the use of cerium oxide-based catalysts for methanol electro oxidation in alkaline medium. A comparison of the electrochemistry of these electrodes in acid and basic conditions is also made.

EXPERIMENTAL

The electrochemical measurements were carried out using a Potentiostat/Galvanostat model 263A EG & G Instruments, Princeton Applied Research with a three electrode cell. The working electrodes were glassy carbon (GC) electrodes from Bioanalytical Systems® (BASi, Indiana, USA). These electrodes were polished with 1.0, 0.3, and 0.05 µm Al_2O_3 paste (Buehler Micropolish®) until a mirror-like state. The residual polishing material was removed from surfaces by sonication of the electrodes in a water bath for 5 min. A platinum foil was used as a counter electrode. Ag/AgCl was used as reference electrode for the measurements in acidic medium. All the other electrochemical data for alkaline medium was obtained with a mercury/mercury oxide (MMO) reference electrode. Solutions were prepared from distilled water that had been treated with Barnstead NANOpure® system giving 18 MX cm water. All other reagents were purchased from Aldrich Chemical Co. The cerium oxide was 99.999% pure and the K_2PtCl_6 salt was 98%. The sulfuric acid was 99.999% pure and the potassium hydroxide salt was 99.95% pure. AFM measurements were carried out using a Nanoscope IIIa-Multimode atomic force microscopy (AFM) from Digital Instruments with a scanning probe microscope controller equipped with a He-Ne laser (638.2 nm) and a scanner type E for AFM experiments.

DISCUSSION

The preparation of Pt/CeO_2 composite electrodes was done according to Cabrera and collaborators [6] using a clean glassy carbon (GC) working electrode. The bath solution contained 20 mM CeO_2 and 1 mM in K_2PtCl_6 in 0.5 M H_2SO_4. The deposition potential was -200 mV vs. Ag/AgCl. The deposition time was 60 s and the composite electrodes were used to get a cyclic Voltammetry (CV) in acid and in alkaline solution. Figure 1 shows two CV's of the composite GC/Pt electrode in 1M methanol (MeOH) in 1M KOH and in 0.5 M H_2SO_4. The

principal difference that is noted is the shift to more negative potentials of the CV's when the electrochemistry is done in alkaline solutions. This result was established by the fact that pH dependant processes show a potential value which is a function of the pH of the solution. Thus, as bigger the pH of the solution, more negative will be the peak potential. The current of the composite electrode was higher in alkaline medium than in acidic medium. Another difference of the peaks from Figure 2 is the fact that in alkaline solution the cathodic peak appears almost at the same potential of the anodic peak. In acidic medium the cathodic and anodic peaks appear at different potentials. This experiment permits conclude that alkaline medium makes more feasible the electro oxidation of methanol than acidic medium. This result was established by Yu and collaborators [8].
In order to determine the size of the particles, electro depositions were done for 30 s on highly ordered pyrolytic graphite (HOPG) using a bath solution of 20 mM CeO_2 and 1mM K_2PtCl_6 in 0.5 M H_2SO_4. Figure 2 is an AFM image that shows the diameter distribution of the particles.

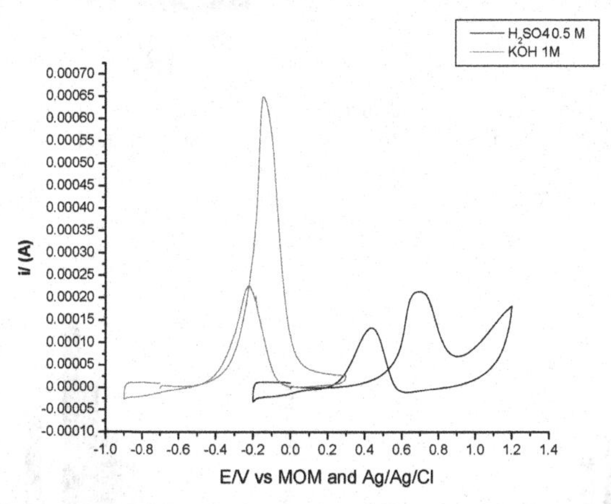

Figure 1. CV of MeOH oxidation in 1M KOH **(left-red)** and 0.5 M H_2SO_4 **(right-black)** of a C/Pt composite electrode prepared for 30 s at -200 mV vs. Ag/AgCl.

Figure 2A shows an AFM image of a clean highly ordered pyrolytic graphite (HOPG) surface and Figure 2B shows the distribution of the Pt/CeO_2 particles after the electro deposition. A statistical analysis was performed in order to find the average size (diameter and height) of the particles.

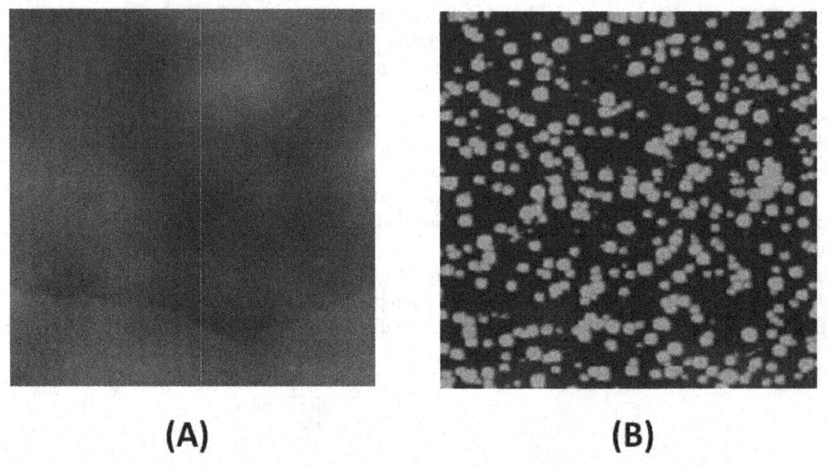

Figure 2. AFM images of a (A) clean HOPG surface and (B) HOPG surface after the modification with 1mM K_2PtCl_6/20 mM CeO_2 for 30 s in 0.5 M of sulfuric acid.
Scan size= 10μmx10μm.

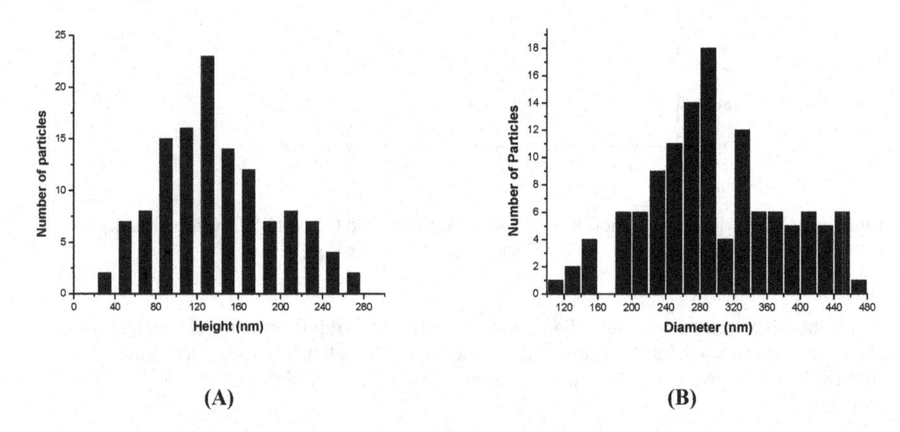

Figure 3. Statistical analysis of the height (A) height and (B) diameter the Pt/CeO_2 particles on HOPG shown in Figure 2B.

From Figure 3A it is possible to deduce that the average height size is 120 nm with a minimum value of 13.5 nm and a maximum of 256 nm. Figure 3B shows the diameter distribution of the Pt/CeO_2 particles on HOPG. In the case of the diameter the average is 258 nm with a minimum value of 76 nm and a maximum value of 437 nm. To determine if that diameter and size of the particles were good enough for methanol electro catalysis, Figure 4 shows a CV in 1M KOH for the composite Pt/CeO_2 and Pt electrodes. Particle size is a function of the preparation method of the particles. As bigger the deposition time of the cerium oxide and K_2PtCl_6 the size of the particles will increase because more material will be deposited on the electrode surface. For this reason, if we want to decrease the particle size, it will be important to make deposits at smaller deposition times (15 s or even 10 s of deposition time). Having smaller particles may allow to have better catalytic response towards methanol electrooxidation because there may be a better distribution and exposed area of the catalysts to the methanol.

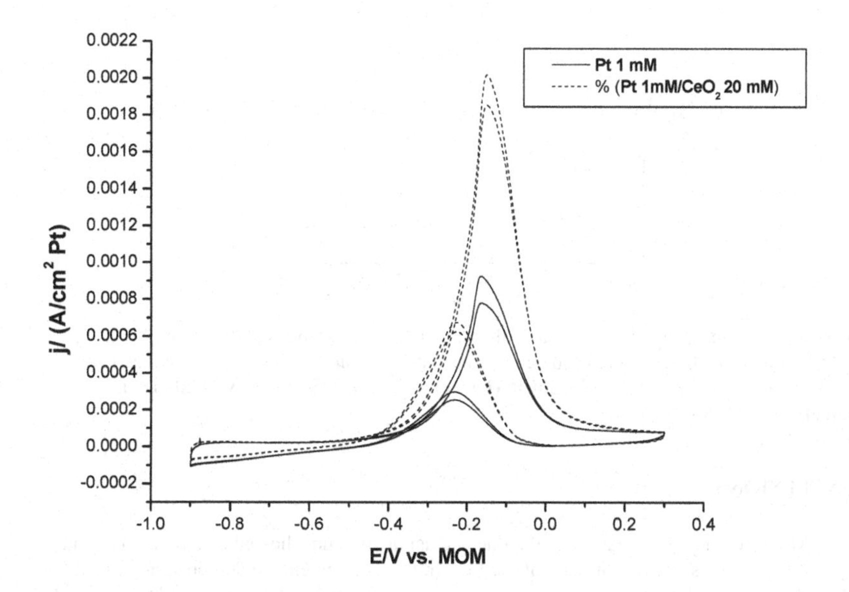

Figure 4. CVs of Pt/CeO_2 and platinum electrodes at 100 mV/s in 1M KOH and 1M CH_3OH. The composites electrodes were prepared at constant potential of -200 mV vs. Ag/AgCl for 60 s from a solution of a)1mM of K_2PtCl_6 and b) 1mM K_2PtCl_6 with 20 mM in CeO_2 .

It can be seem that the only Pt electrode showed potential shifts towards more positive values than the Pt/CeO_2 composite electrode. This is indicative that the latter has an enhancement on the electro oxidation of methanol. The currents on both CV's were normalized using the adsorption-desorption area peaks of the CV's in 0.5M sulfuric acid. The current density of the CV for methanol oxidation is bigger than the one for the only platinum electrode . In order to proof that, chronoamperometric responses were done for those two electrodes and the results are shown in Figure 5. As it is observed, the catalytic activity for the Pt/CeO_2 electrode was superior to the one for the Pt-only electrode.

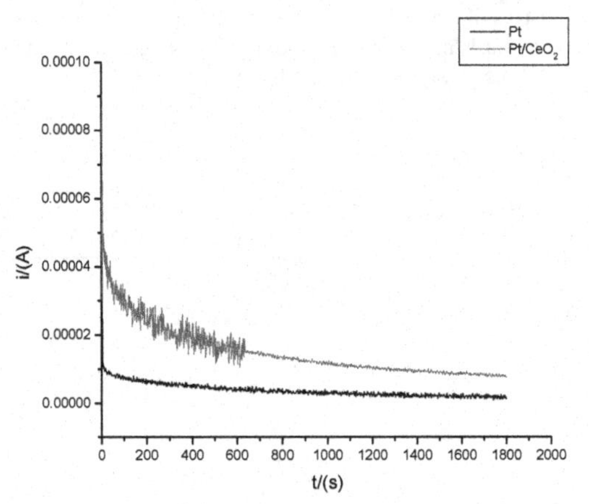

Figure 5. Chronoamperometric response of two composite electrodes at -0.4 V vs. MOM. The composites electrodes were prepared at constant potential of -200 mV vs. Ag/AgCl for 60 s from a solution of a)1mM of K_2PtCl6 **(lower)** and b) 1mM K_2PtCl_6 with 20 mM in CeO_2 **(upper).**

CONCLUSIONS

Methanol electro oxidation in alkaline and acidic medium showed differences beyond the shift of the potentials of the oxidation of the fuel. In this experiment we demonstrated that the oxidation current is larger in alkaline medium than in acidic medium for the same catalysts. The catalyst with cerium oxide showed a better current than the platinum catalysts under the same conditions. It is recommended to use different conditions to prepare the catalysts and evaluate the catalytic currents.

ACKNOWLEDGMENTS

The authors gratefully acknowledge the financial support from the Puerto Rico Institute for Functional Nanomaterials NSF- EPSCoR-IFN Grant No. OIA-070152, 1002410 and the Center for Advanced Nanoscale Materials NASA-URC Grant Number NNX10AQ17A.

REFERENCES

1) Z.X. Liang, T.S. Zhao, J.B. Xu and L.D. Zh, Electrochim. Acta. 54, 2203 (2009).

2) J. Bagchi and S. K. Bhattacharya , J. of Power Sources. 163, 661 (2007).

3) M.A. Abdel Rahim, R.M. Abdel Hameed and M.W. Khalil, J. of Power Sources 134, 160 (2004).

4) A. Abdul, A. Khan, A. Riaz and L. M. Muhammad, Turk J. Chem 32, 743 (2008).

5) J. Wang, J. Xi, Y. Bai , Y. Shen, J Sun, L. Chen, W. Zhua and X. Qiu, J. of Power Sorces 164, 555 (2007).

6) C.L. Campos, C. Roldán, M. Aponte, Y. Ishikawa and C.R. Cabrera, J. of Electroanal. Chemistry 581, 206 (2005).

7) D. Diaz, N. Greenletch, A. Solanki, A. Karakoti and S. Seal, Catal. Lett. 119, 319 (2007).

8) E. Yu, K. Scott and R. Reeve, J. of Electroanal. Chem.547, 17 (2003).

Mater. Res. Soc. Symp. Proc. Vol. 1446 © 2012 Materials Research Society
DOI: 10.1557/opl.2012.1107

In situ spectroscopic characterization of some $LaNi_{1-x}Co_xO_3$ perovskite catalysts active for CH_4 reforming reactions

Rosa Pereñiguez, Victor M. Gonzalez-Delacruz, Fatima Ternero, Juan P. Holgado, Alfonso Caballero
Instituto de Ciencia de Materiales de Sevilla and Dep. Química Inorgánica (CSIC-Universidad de Sevilla). Avenida Américo Vespucio, 49, 41092 Sevilla (España)

ABSTRACT

Lanthana-supported Ni and Co catalysts were investigated by "operando" techniques (XAS and APPES) for methane reforming reactions. The samples were prepared by the "solid phase crystallization" method (spc), where the precursors $La(Ni_{1-x}Co_x)O3$ contains homogeneously distributed metals (Ni, Co) in the crystal structure (perovskite), which, on further reduction, result in the formation of catalytic system $Ni_{1-x}Co_x/La_2O_3$. The monometallic samples ($NiLaO_3$, $CoLaO_3$) have been compared with a bimetallic system of an intermediate composition $Ni_{0.5}Co_{0.5}LaO_3$. This "operando" study has allowed us to obtain important conclusions about the bimetallic particles and the metal-support interactions. The data revealed the formation of bimetallic particles (NiCo); on these ones, the Ni avoids the Co oxidation during the reaction. However, this protection does not induce an improvement in the activity, which presents an intermediate behaviour between Ni/La_2O_3 and Co/La_2O_3. These bimetallic particles form a pseudo-alloy with the surface enriched in cobalt (under reduced conditions), resulting nearly in a core-shell structure (Ni@Co).

INTRODUCTION

The dry reforming of methane has awaked a great interest in the last years versus the traditional steam reforming, especially for the syngas ($CO+H_2$) production. This process offers important advantages compared to steam reforming of methane. In this context, the low H_2/CO ratios obtained by dry reforming, makes this process interesting for the liquid hydrocarbons production (Fischer Tropsch reaction) and for the production of others products as formaldehyde, polycarbonates or methanol. In general, this process is not yet industrialized, but it should be remarked that the high concentration of CO_2, in the natural gas yield and in the biomass for fermentation (ca. 30-60%), together with the methane, makes this process attractive to the industrialization. However, the main problems are the coke formation and the high temperatures (ca. 800 °C) required for the reaction. There are several factors under study for understanding the carbon formation as the support used or the temperature of reduction, but the principal investigation is focused on the metallic phase. Although the good results obtained for the noble metals, their high cost makes necessary the searching of cheaper alternative, like nickel or cobalt due to their inherent availability, low cost and high activity. However, these metals present also some disadvantages: the Ni produces an important coke deposit; while the Co is less resistant to oxidize in the reaction conditions [1]. In fact, some studies are directed to prevent the Co oxidation through the addition of noble metals or an alloy formation as NiCo. In this context, we have focused the study to bimetallic Ni-Co systems supported on La_2O_3 prepared from the reduction of the precursor $LaNi_{1-x}Co_xO_3$ (by solid phase crystallization) with a perovskite structure. It is worthy to stand out that, there is still a lot of controversy on aspects such as the nature/state of the active phase under "operando" conditions, in spite of the effort made in the investigation of these systems.

EXPERIMENTAL

The $LaNi_{1-x}Co_xO_3$ perovskites were prepared by the spray pyrolysis method [2], using a solution of $La(NO_3)_3$ and $Ni(NO_3)_2/Co(NO_3)_2$, which is passed through two on-line furnaces at 250°C and 600°C respectively, producing a powder that was later calcined in air at 600°C. The physicochemical state of the powders was characterized by means of SEM, XRD, TPR, XPS, etc. The measurements of the catalytic performance in the DRM/SRM reactions were accomplished using an atmospheric flow reactor. The feed gas was a methane/carbon dioxide (or water)/helium mixture with a space velocity of 300000 ml/hg. XAS spectra were collected in transmission mode at the BM25 station of the ESRF (Grenoble, France), while the Ambient Pressure Photoemission Spectroscopy (APPES) experiments were performed at beam line U49/2-PGM1 at BESSY II (Berlin, Germany).

RESULTS AND DISCUSSION

The catalytic performances of the perovskite samples have been determined for the dry and steam reforming of methane. From these results, it can be concluded that these catalysts are more stable for the dry reforming reaction, as the catalysts deactivate very quickly under steam reforming reaction conditions. So, while under dry reforming conditions the activity remains stable after 10 hours, a decrease of 20% in the methane conversion occurs under steam reforming reaction conditions. On the other hand, the catalytic activity is higher as the proportion of nickel increases: 5% for the cobalt monometallic sample, 40% for the bimetallic one, reaching values of 90% conversion for the monometallic $LaNiO_3$ catalyst at 800°C.

The XAS spectra obtained for the different samples at the Ni and Co K edges are presented in Figure 1. As shown, the nickel phase in the reduced $LaNiO_3$ and $LaNi_{0.5}Co_{0.5}O_3$ samples, analyzed by operando XAS, evolves from a mixture of Ni^{3+} and Ni^{2+} to metallic Ni^0 under both, hydrogen reduction treatment and DRM reaction. This behavior contrasts with the partial oxidation of nickel observed under SRM reaction [3].

The results obtained by in situ XAS for Co phase for DRM and SRM in the $LaCoO_3$ and $LaNi_{0.5}Co_{0.5}O_3$ reveal that in both cases the cobalt phase is partially oxidized, remaining visibly less oxidized in the $LaNi_{0.5}Co_{0.5}O_3$ that in the monometallic catalysts. These results could be explained considering the formation of a NiCo bimetallic alloy after hydrogen reduction of the original Ni-Co perovskite.

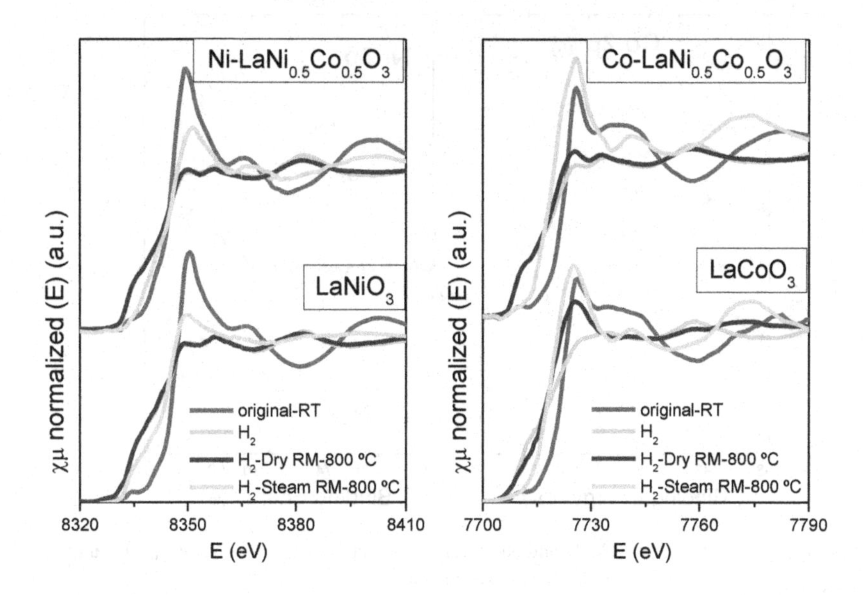

Figure 1 - Ni K edge (left) and Co K edge (right XANES spectra obtained in situ for the Ni, Co and NiCo perovskites.

The APPES spectra obtained for the $LaNi_{0.5}Co_{0.5}O_3$ sample submitted to a hydrogen reduction treatment agree with the XAS results. As shown in Figure 2, the XPS data indicate that both metallic phases (Ni and Co) remain reduced after hydrogen treatment and dry reforming reaction, while a slight oxidation process occurs under steam reforming reaction conditions, where both metals are partially oxidized to Ni(II) and Co(II) respectively.

Even more interestingly, the XPS spectra obtained for this sample with different incident photon energies, corresponding to kinetic energies (KE) of the photoelectron of 200 and 600 eV, shows an important decrease in the intensity of the Ni 3p signal with 200eV photoelectrons. This finding allows us to propose a structure for the bimetallic particles, where the metals are arranged as a "pseudo core-shell" Ni@Co.

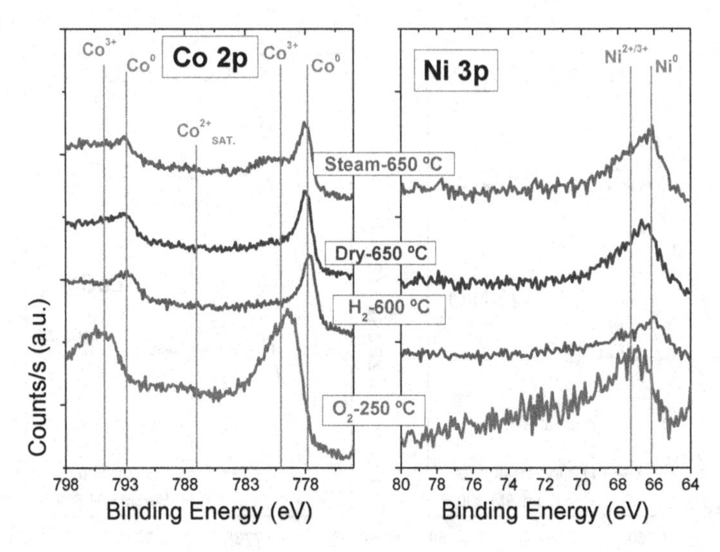

Figure 2 - APPES spectra (Co 2p and Ni 3p regions) obtained for the $LaNi_{0.5}Co_{0.5}O_3$ sample submitted to the indicated treatments.

CONCLUSIONS

The catalytic tests revealed that the catalysts based on nickel are more active and selective for the DRM reaction, meanwhile the presence of cobalt results in a lower activity and the enhancement of the RWGS. Besides, the studied systems present better results for the DRM than for the SRM.

The nickel phase behaviour in the $LaNiO_3$ and $LaNi_{0.5}Co_{0.5}O_3$ analyzed by operando XAS evolves from the mixture of $Ni3^+ + Ni^{2+}$ in the original samples, to Ni^0 under reducing conditions and DRM reaction, in contrast with the partial oxidation registered in SRM reaction. The effects detected in the cobalt phase for DRM and SRM in the $LaCoO_3$ and $LaNi_{0.5}Co_{0.5}O_3$ reveal a bigger oxidability in the cobalt phase than the observed for the nickel species, and this oxidability is lower in the $LaNi_{0.5}Co_{0.5}O_3$ sample. We propose that this higher resistance to oxidation in the mixed sample is promoted by the nickel phase and is transferred to the cobalt phase due to the formation of bimetallic alloy of NiCo, which has been previously observed in similar systems by other authors (4-6). The operando APPES data for $LaNi_{0.5}Co_{0.5}O_3$ agree with the XAS results, detecting this slight oxidation under the SRM reaction. Registering the spectrum of $LaNi_{0.5}Co_{0.5}O_3$ at H2-600 °C using different photon energies (200 and 600 eV) help us to propose a possible structure for the bimetallic alloy where the metals are arranged as a "pseudo core-shell" Ni@Co..

ACKNOWLEDGMENTS

We thank the Ministry of Education and Science of Spain and Junta de Andalucía for financial support (Projects ENE2011-24412 and P07-FQM-02520), the ESRF and BESSY II facilities and the BM25 Spline beamline staff for their experimental support.

REFERENCES

1. K. Takanabe, K. Nagaoka, K. Nariai, K. Aika; J. Catal., 232 (2005) 268.
2. E. López-Navarrete, M. Ocaña; J. Europ. Cer. Soc., 22 (2002) 353-359.
3. R. Pereñíguez, V.M. González-DelaCruz, J.P. Holgado, A. Caballero; Appl. Catal. B: Env., 93 (2010) 346-353.
4. V.M. González-DelaCruz, R. Pereñíguez, F. Ternero, J.P. Holgado, A. Caballero; J. Phys. Chem. C, 116 (2012) 2919.
5. J. Zhang, H. Wang, A.K. Dalai, J. Catal. 249 (2007) 300.
6. K. Nagaoka, K. Takanabe, K. Aika, Appl. Catal. A 268 (2004) 151.

Mater. Res. Soc. Symp. Proc. Vol. 1446 © 2012 Materials Research Society
DOI: 10.1557/opl.2012.954

Methane Combustion Using CeO_2-CuO Fibers Catalysts

Felipe A. Berutti[1], Raquel P. Reolon[1], Annelise K. Alves[1], and Carlos P. Bergmann[1]
[1]Federal University of Rio Grande do Sul, Av. Osvaldo Aranha, 99 sl. 705C,
Porto Alegre, RS, 90035190, Brazil

ABSTRACT

The use of three-way catalysts is an accepted method to minimize NOx and CO emissions generated by internal combustion engines. These catalysts are generally formed by the support, stabilizers, promoters metal and transition metals, the most used metals of the platinum group. The use of cerium as a promoter is usually related to its ability to store oxygen and structural aspects such as the property of increasing the dispersion of metals and slow change of phase of the stabilizing support. On the other hand, the metal copper was explored as a possible replacement for palladium and platinum in the reduction of NO by CO. In this work, fibers of cerium oxide doped with copper were obtained from an acetate solution of cerium and coppers nitrates and polyvinyl butyral (PVB). This solution went through the process of electrospinning to produce nanostructured fibers. After heat treatment, cerium oxide fibers were obtained. These fibers were characterized structurally by scanning electron microscopy (SEM), had their specific surface area determined by BET method, were subjected to thermogravimetric test to determine their thermal decomposition and were analyzed by X-ray diffraction. The catalytic activity was assessed by the amount of O_2 consumed and CO and CO_2 formed for the combustion of methane and air. SEM images show fibers oriented randomly in the substrate. TEM images show that the diameter of the fibers is approximately 100 nm and the size of its crystallites are around 20 nm. In the presence of the catalyst, the combustion reaction started around 500°C, with the consumption of methane and oxygen and the formation of CO and/or CO_2. There was no emission of NO and NOx gases during the tests with catalysts.

INTRODUCTION

The interest in the study and application of nanoparticles has been growing in recent years, mainly due to its unique physical and chemical properties that make them significantly different from their usual microstructure [1]. Nanostructured particles have high catalytic activity due to their high surface area and surface properties (such as surface defects) [2].

The electrospinning technique has been recognized as a versatile and effective method for producing fibers with very small diameters and high surface-volume ratio [3]. The morphology and properties of fibers depend on the characteristics of the polymer and process parameters used, for example, average molecular weight of polymer, solvent, viscosity and conductivity of the solutions, applied electric field strength and distance from the collector [4,5,6].

The use of three-way catalysts is an accepted method to minimize NOx and CO emissions generated by internal combustion engines. These catalysts are generally formed by the support, stabilizers, promoters metal and transition metals, the most used metals of the platinum group [7].

The use of cerium as a promoter is usually related to its ability to store oxygen [8] and structural aspects such as the property of increasing the dispersion of metals and slow change of phase of the stabilizing support.

On the other hand, the metal copper was explored as a possible replacement for palladium and platinum in the reduction of NO by CO [9]. Despite the importance of oxidation of CO on Cu, the reaction is not elucidated because changes in oxidation state when the reaction conditions are changed. Somorjai and coworkers [10] studied the catalytic activity of Cu^0, Cu^+ and Cu^{+2}. The results indicated that the catalytic activity for CO oxidation decreased from Cu^0 to Cu^{+2} and was inhibited by oxygen. The type of mechanism was explained in the Langmuir-Hinshelwood where adsorbed CO reacts with adsorbed oxygen. Oxygen should be decoupled in order to have the reaction. For Cu^{+1}, the higher activation energy found in relation to Cu^0 was explained as extra energy for dissociation of O_2.

In this paper, the fibers of cerium oxide doped with copper were obtained from an acetate solution of cerium nitrate, copper and polyvinyl (PVB). This solution went through the process of electrospinning to produce nanostructured fibers. After heat treatment, cerium oxide fibers were obtained. These fibers were characterized structurally by scanning electron microscopy (SEM), had their specific surface area determined by BET, were subjected to thermogravimetric test to determine their thermal decomposition and were analyzed by X-ray diffraction. The catalytic activity was assessed by the amount of O_2 consumed and CO and CO_2 formed for the combustion of methane and air.

EXPERIMENTAL

A solution containing cerium nitrate, determined amounts of copper nitrate and anhydrous ethyl alcohol was prepared. This solution was mixed with 12ml of a solution containing 15 wt% of polyvinyl alcohol (PVB, Clariant) in ethanol. The final solution was used in the electrospinning process immediately after preparation.

To measure the catalytic activity of this material after heat treatment it was used a muffle furnace with a quartz tube mounted vertically inside. The catalyst (0.29 g) and a flow of methane (99.995% purity) of 0.1 L / min and synthetic air (20% O_2 and 80% N_2) of 0.9 L / min were used. The amount of gas CxHy, O_2, CO, CO_2, NO, NOx were measured using a portable gas analyzer model Ecoline 4000 (Eurotron, Italy).

Electrospinning

Fibers were obtained using the electrospinning technique from the mixture of cerium and copper nitrates and PVB. In a typical electrospinning process, the precursor solution is placed in a 5 ml syringe connected to a 10-gauge hypodermic needle. The needle is connected to a high voltage source. The voltage of 15 kV was used, applied at a separation distance of 12 cm between the needle and a cylindrical counter-electrode covered with aluminum foil. The flow of the fluid is controlled by an infusion pump, and kept constant at 0.8 mL/h.

Characterization

The crystalline phases were identified through X-ray diffraction analysis using a Philips equipment (model X'Pert MPD) operating at 40 kV and 40 mA with Cu K_α radiation. The analysis was performed at a rate of 0.05°/min with a step of 1 sec at a range of 5 to 75°.

The morphology of the powder produced in this work was observed by scanning electron microscopy (SEM) using a Jeol equipment (model JSM 5800). A thin gold layer was deposited on the sample surface to make it conductive.

Thermogravimetric analysis (SDTA 851 Metler & Toledo) was performed at 1000°C with a heating rate of 10°C/min in an atmosphere of synthetic air.

The specific surface area (Nova 1000, Quantachrome) was determined using the BET (Brunauer-Emmett-Teller) method with N_2 as adsorbent. The samples were previously kept in vacuum and temperature of 70 °C for 3 hours.

RESULTS AND DISCUSSION

The thermal decomposition of fibers containing cerium and copper nitrates and PVB was accompanied by thermal analysis, as shown in Figure 1.

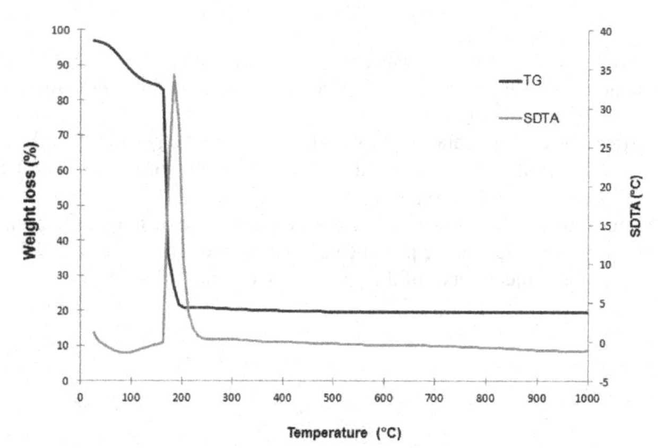

Figure 1 - Thermal analysis of fibers containing cerium, copper and PVB.

The thermogravimetric analysis curve shows a continuous mass loss up to approximately 220 °C, probably caused by loss of ethanol, decomposition of PVB and nitrates. Above 300°C there are no further significant mass losses. The differential thermal analysis curve shows an exothermic peak around 200°C associated with a mass loss probably related to the degradation of PVB.

Figure 2a shows an image obtained by SEM and Figure 2b shows an image obtained by TEM of the fibers after heat treatment at 650°C.

The average fiber diameter is 1 μm after the heat treatment. If we compare the diameters before and after heat treatment there is a reduction of the diameter of the fibers caused mainly by the loss of organic compounds during heat treatment. The average size of crystallites, observed by transmission electron microscopy is about 25nm. The specific surface area of the heat treated fibers was about 49.7 m^2/g.

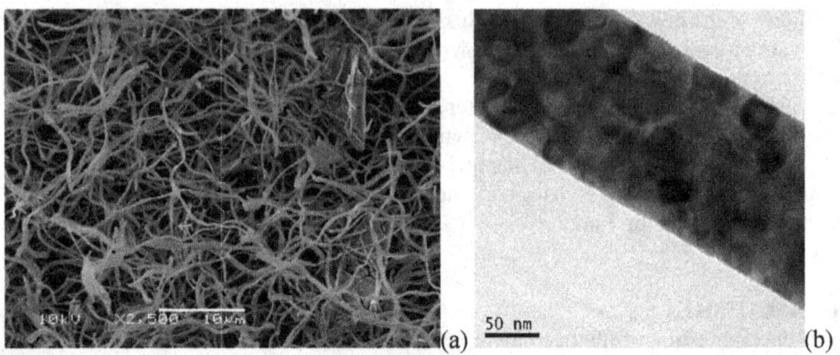

Figure 2 - Electrospinning fibers obtained after heat treatment 650°C: (a) SEM and (b) TEM images.

The catalytic activity of the fibers was measured by analyzing the gases consumed and generated during a combustion test. The material was heated up to 600°C in a gas flow mixture containing 10% methane and 90% synthetic air.

A first test was carried out without catalyst, by simple heating of the gas mixture. The quantities of gases CxHy and O_2 remained constant and it was detected the formation of NO and NOx up to 4ppm during this test. No combustion occur at this condition.

Further tests were made using the synthesized catalysts. Figure 3 represents a diagram presenting the results of a combustion test in the presence of catalyst (cerium oxide with 0.5 wt% of copper), showing the variation in the amount of the gases O_2, CO_2 and CxHy.

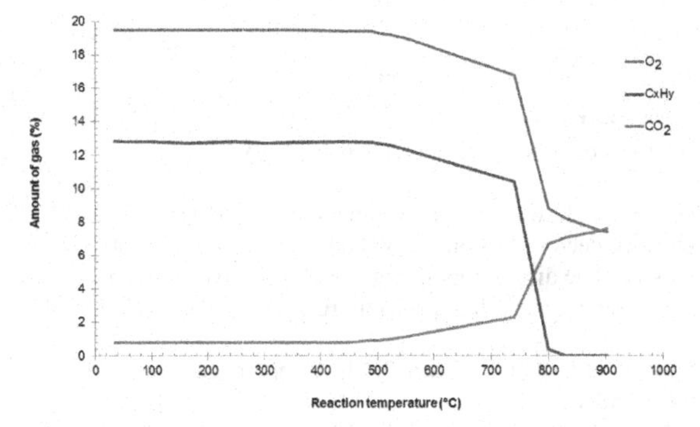

Figure 3 - Variation in the concentration of the gases O_2, CO_2 and CxHy during the process of methane combustion using fibers containing 0.5 wt% of copper.

In the presence of 0.29 g of catalyst synthesized by electrospinning, the fibers containing cerium oxide and 0.5 wt% copper initiate the combustion process around 550°C, when it was noticed a significant decrease in the amount of CxHy and O_2 and an increase in the amounts of

CO_2. The formation of CO was not detected during the catalytic tests using 0.5 wt% of copper in the catalyst. The values of methane concentrations have fallen below 1 %, indicating almost complete conversion of this gas. The catalysts containing 0, 1 and 2.5 wt% of copper had similar catalytic behavior, except concerning the formation of CO that was detected in concentrations below 0.2 %.

CONCLUSIONS

Fibers of cerium oxide and copper were obtained by electrospinning technique. SEM images show fibers oriented randomly in the substrate. TEM images show that the diameter of the fibers was approximately 100 nm and the size of its crystallites was around 20 nm.

The catalytic activity of the fibers was significant. In the absence of catalyst up to 600 °C there was no combustion reaction of methane and air mixtures. In the presence of the catalyst, the combustion reaction started around 550°C, with the consumption of methane and oxygen and the formation of CO and/or CO_2. In the presence of the catalyst there was no emission of NO and NOx gases. The best results were obtained when using cerium oxide doped with 0.5 wt% of copper, when the methane conversion was complete and the ignition temperature 510 °C.

REFERENCES

[1] Tok, A.I.Y.; Boey, F.Y.C.; Du, S.W.; Wong, B.K. Mat. Sci. Eng. B-Solid., v.130, p.114-119, 2006.
[2] Hwang, C.-C.; Wu, T.-Y;. Mat. Sci. Eng. B-Solid., v.111, p.197-206, 2004.
[3] Reneker D H and Chun I 1996 Nanotechnology 7 216.
[4] Reneker D H, Yarin A L, Fong H, Koombhongse S 2000 J. Appl. Phys. 87 4531.
[5] Deitzel J M, Kleinmeyer J, Harris D and Beck T N C 2001 Polymer 42 261.
[6] Shin Y M, Hohman M M, Brenner M P and Rutledge G C 2001 Polymer 42 9955.
[7] Harrison, B.; Diwell, A. F.; Hallet, C.; Platinum Metal Rev. 1988, 32, 73.
[8] Serre, C.; Garin, F.; Belot, G.; Maire, G.; J. Catal. 1993, 141, 1.
[9] Petrov, L. A.; Soria, J.; Conesa, J. C.; Coronado, J. M.; Martínez-Arias, A.; Cataluña, R.; Arcoya, A.; Seoane X. L.; Catal. and Aut. Pol. Control III 1995, 96, 215.
[10] Somorjai, G. A.; Jernigan, G. G.; J. Catal. 1994, 147, 567.

Mater. Res. Soc. Symp. Proc. Vol. 1446 © 2012 Materials Research Society
DOI: 10.1557/opl.2012.1221

Role of Surface Oxide Layer during CO_2 Reduction at Copper Electrodes

Cheng-Chun Tsai, Joel Bugayong, and Gregory L. Griffin

Department of Chemical Engineering, Louisiana State University, Baton Rouge, LA 70803, U.S.A.

ABSTRACT

We have compared the rates of CO formation on Cu and Cu oxide surfaces during the electrochemical reduction of CO_2 in aqueous media. On metallic Cu surfaces, H_2 formation is the main reaction at potentials less cathodic than –1.16 V(NHE). At this potential the formation of CO becomes significant, while CH_4 appears at potentials more cathodic than –1.36 V(NHE). On electrodeposited Cu oxide surfaces there is a complex transient response. During reduction at constant potential (–1.1 V(NHE)), there is a large, transient cathodic current that corresponds to reduction of the oxide layer. After this initial oxide reduction, the current density stabilizes and the formation rates of H_2 and CO show a more slowly varying transient behavior. The H_2 formation rate is roughly 3x higher than on freshly cleaned Cu foil, but is largely independent of the thickness of the initial oxide layer. In contrast, the CO formation rate is at least one order of magnitude higher on the (reduced) Cu oxide samples than on Cu foil at the same potential. These results are interpreted as evidence that CO formation is enhanced at low-coordination number Cu sites present on freshly nucleated Cu clusters following oxide reduction.

INTRODUCTION

The electrochemical reduction of CO_2 to produce chemicals and liquid fuels has been identified as a priority research direction and a basic research need by the U.S. Department of Energy [1]. Global consumption of liquid fossil fuels is currently around 80 million barrels (oil equivalent)/day, and is projected to rise to 100 million barrels/day by 2030 [2]. The current value corresponds to a CO_2 release rate into the atmosphere of 3.6 Gt C/year [3]. For comparison, the net consumption of CO_2 by terrestrial biological sinks (i.e., plant life, including land use changes) is estimated to be 1.1 GtC/year [4]. Thus it is apparent that an alternative, sustainable method of producing hydrocarbon fuels will be necessary if these fuels are to remain a viable long-term option for society.

Copper is widely recognized for its unique selectivity for producing hydrocarbon products during electroreduction of CO_2 in aqueous media [5, 6]. Low H_2 overpotential metals (e.g., Pt, Ni, Fe) primarily form H_2; these metals are also considered to bind CO too strongly to form any other products. High overpotential metals (Hg, Cd, Pb, Tl, In, Sn) cannot bind CO_2 strongly enough for dissociative reactions, and instead produce HCOOH as the major CO_2 reduction product. Medium overpotential metals (Ag, Au, Zn) can dissociate CO_2, but they do not bind CO strongly and release it as their primary product (although this may be of interest for the electrochemical production of H_2/CO syngas mixtures). Copper alone has the ability to form and retain adsorbed CO for subsequent electrochemical hydrogenation.

The oxidation state of the Cu electrode has been suggested to play a role in the product selectivity. Most authors have reported the formation of light hydrocarbons, notably CH_4 and C_2H_4 (but not C_2H_6), when CO_2 is electrochemically reduced at Cu metal surfaces. In contrast, Frese reported high efficiencies for CH_3OH formation using oxidized Cu electrodes [7]. (The CH_3OH product would be more useful than light hydrocarbons for energy storage because of its higher energy density.) Chang *et al.* reported CH_3OH formation using Cu_2O nanoparticles encapsulated in Nafion film electrodes [8]. Other members of our Center have reported high CH_3OH efficiency using model Cu/ZnO electrodes [9]. However, operating stability and the reproducibility of sample preparation were significant concerns in the latter report [9].

This concern for the stability of Cu oxide electrodes arises because the standard potentials for the reduction of both CuO and Cu_2O are less cathodic than the potentials for reduction of CO_2 [5]:

$$CuO + 2H^+ + 2e^- = Cu + H_2O \qquad E_0 = -0.127\ V(NHE)$$
$$Cu_2O + 2H^+ + 2e^- = 2Cu + H_2O \qquad E_0 = -0.052\ V$$

vs.

$$CO_2 + 2H^+ + 2e^- = CO + H_2O \qquad E_0 = -0.521\ V$$
$$CO_2 + 6H^+ + 6e^- = CH_3OH + H_2O \qquad E_0 = -0.409\ V$$

Thus Cu oxide surfaces are thermodynamically unstable under CO_2 reduction conditions and will eventually be reduced to Cu metal.

In this work we compare the rate of CO formation (i.e., the initial step in CO_2 reduction) using clean and oxide-covered Cu electrodes. This is intended as a first step toward developing catalysts where the CuOx active phase can be stabilized (e.g., by dispersion on an interactive support, or by partial isolation using a protective matrix). The Cu oxide layers are prepared using electrochemical deposition onto Cu substrates. The rate of CO formation is measured as a function of initial oxide layer thickness and time on stream, and compared with the rate observed on Cu metal surfaces.

EXPERIMENT

All samples were prepared from copper foil that was cut into 1cm^2 pieces. Metallic Cu surfaces were prepared using the foil as received, after anodic electropolishing in 85% H_3PO_4 solution at 0.59 V(NHE) until shiny. Copper oxide samples were prepared by electrochemical depositing Cu_2O from a bath containing 0.4M copper sulfate and 3.0M lactic acid at – 0.31 V (NHE) and 65°C for 30 minutes. The bath was adjusted to pH = 9.0 by addition of NaOH pellets. The oxide deposition step was performed using a dedicated one-compartment cell, separate from the two-compartment cell used for CO_2 reduction measurements.

All experiments were controlled using a Princeton Applied Research Model 263A potentiostat. The electrochemical reactor was a two compartment cell, using a Nafion membrane separator, a Pt wire counterelectrode, and Ag/AgCl reference electrode (working electrode potentials are reported here as V(NHE)). The electrolyte volume in each side of the cell was 20 mL. The CO_2 electrochemical reduction was performed using a solution of 0.5M $KHCO_3$ solution pre-saturated with CO_2 for 30 minutes. During the run, the electrolyte was also bubbled with CO_2 at a rate of 40 mL/min.

The concentration of each species in the CO_2 gas flow leaving the reactor was determined using gas chromatography. The measured reaction products included H_2, CO, and CH_4. The rate of formation of each product (μmole/cm^2/hr) was determined by combining the product concentration, the exit gas flow rate, and the superficial surface area of the electrode. Faradaic efficiencies were determined by comparing the rate of product formation and the measured current density.

RESULTS

All of the results presented here were obtained in the region of cathodic potential corresponding to the onset of CO formation, in order to reduce the competing effect of subsequent CO conversion to more highly reduced products. (This is reflected in the low CH_4 formation rates listed in the Table 1.) Separate experiments confirmed an increase in CH_4 formation and the onset of C_2H_4 formation if the electrode is operated at higher cathodic potentials.

Product formation on Cu metal surfaces

Product formation rates observed using freshly cleaned Cu foils are shown in Table 1. Rates are obtained for electrode potentials of –0.96 V(NHE), –1.16 V(NHE), and –1.36 V(NHE). The current density increases from 0.5 mA/cm^2 to 4 mA/cm^2 over this range, in reasonable agreement with current densities reported by previous authors [10]. The H_2 generation rate ranges from 5 - 65 μmole/cm^2/hr, corresponding to Faradaic efficiencies approaching 100%. Formation of CO is found at –1.16 V(NHE) and –1.36 V(NHE). The CO formation rate increases from 0.03 to 0.23 μmole / cm^2 / hr, which corresponds to Faradaic efficiencies of 0.2 - 0.4% . Methane is observed only at –1.36 V(NHE), with current density of 0.05 μmole/cm^2/hr and Faradaic efficiency around 0.3%.

Table 1: Product formation rates using clean (non-oxidized) Cu metal surfaces (μmole/cm^2-h).

	Potential (V (NHE))		
	-0.96V	-1.16V	-1.36V
H_2	5.75±2.82	15.07±2.59	64.40±7.59
CO	N.D.	0.03±0.01	0.23±0.05
CH_4	N.D.	N.D.	0.05±0.01

Electrodeposition of Cu oxide layer

We next examined the reduction behavior of the ECD Cu_2O films deposited on Cu electrodes. The CO_2 reduction is performed at –1.1 V(NHE). The transient current vs. time response at this potential is shown in Figure 1. Significant reduction current is observed immediately and persists for one or two minutes. The integrated area under the current vs. time curve correlates with the deposition time used for film growth. The initial reduction process is accompanied by a color change of the working electrode from purple to reddish matte. Taken together, these results indicate that the initial phase of the electrode response corresponds to the nearly complete reduction of the Cu oxide film.

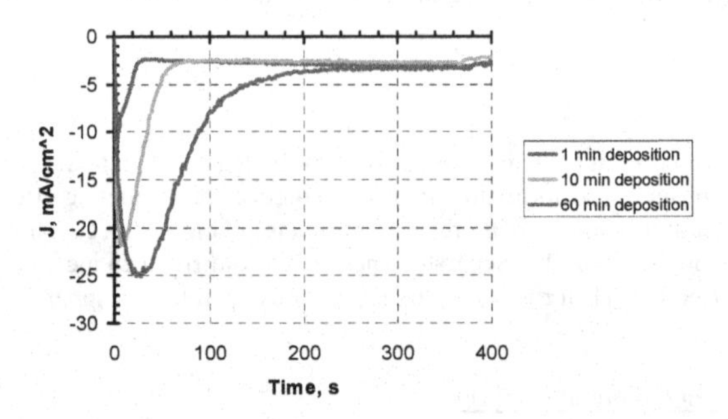

Figure 1: Transient initial current response during CO_2 reduction on Cu_2O films prepared using different deposition times.

Product formation on Cu oxide layer

Following the decay of the oxide reduction current noted above, we continue to hold the electrode at the same potential and monitor product formation (cf. Figure 1). The current density is stable around 3 mA/cm^2 and does not decay significantly during the15 minutes between successive GC sampling times. The current density does not depend on the thickness of the initial ECD oxide layer. The value is roughly 2x higher than observed for Cu foil samples. We interpret this to indicate that the ECD oxide film has a higher surface roughness, but that the roughness factor is independent of layer thickness.

As shown in Figure 2, the H_2 production rate is largely independent of the initial Cu_2O layer thickness and shows an increase with time on stream. For the thinnest oxide layer, the

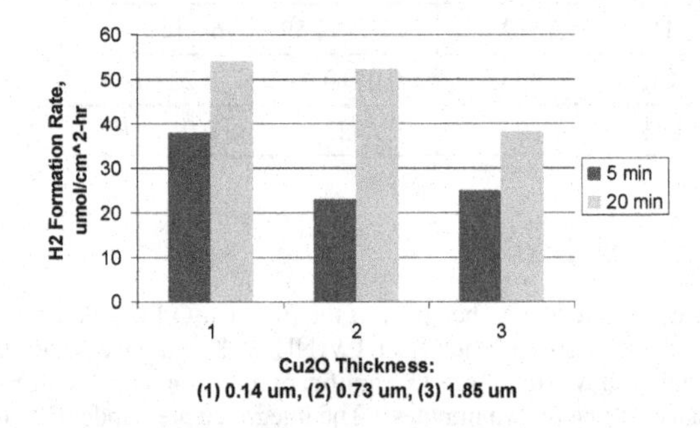

Figure 2: Influence of initial Cu_2O layer thickness on H_2 formation rate.

production rate reached 38 μmole/cm^2/hr after 5 minutes and 54 μmole/cm^2/hr after 20 minutes; these rates correspond to Faradaic efficiencies of 67% and 98%, respectively. For the intermediate thickness layer, the H_2 production rate was only 23 μmole/cm^2/hr after 5 minutes (F.E. = 43%), but reached 52 μmole/cm^2/hr after 20 minutes. For the thickest sample, the H_2 rate after 5 minute was similar (25 μmole/cm^2/hr), but only increased to about 40 μmole/cm^2/hr after 20 minutes.

In contrast, the CO formation rate increases with thickness and decreases with time on stream (cf. Figure 3). The highest rate observed approaches 6 μmole/cm^2/hr for the thickest oxide sample at the end of the first 5 minute sampling interval. This corresponds to a Faradaic Efficiency of 10%. The CO formation rates are more than 1 order of magnitude greater than observed for clean Cu foil. This is much larger than the differences observed for current density and H_2 production rate that were attributed to surface roughness. Instead, it indicates there is a significant change in the product selectivity on the oxide-derived surfaces.

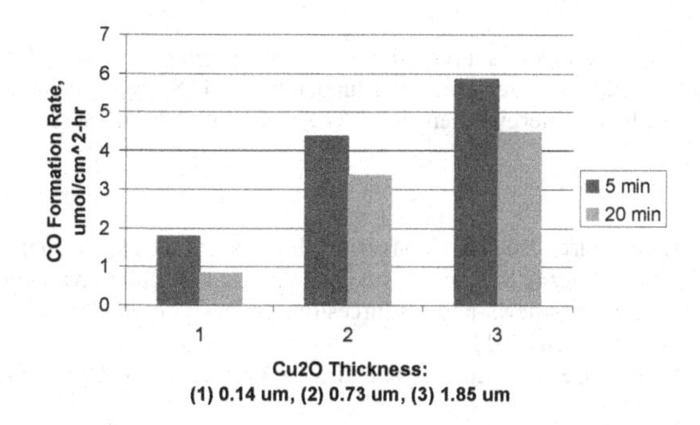

Figure 3: Influence of initial Cu_2O layer thickness on CO formation rate.

DISCUSSION

The electrochemical reduction of CO_2 is a thermodynamically challenging reaction [11]. Theoretical calculations using the Cu(211) surface indicate the reaction proceeds via an adsorbed $COOH_{(a)}$ intermediate, which is subsequently reduced to form CO and H_2O [12]. Preliminary calculations by other members of our EFRC center have confirmed the favorability of this pathway, and have further shown that the free energy of formation for the COOH(a) intermediate is reduced on the high index Cu surface (e.g., Cu(211) as compared to the Cu(111) surface). Additional calculations indicate the energy of formation is reduced even further on various Cu oxide surfaces (e.g., on Cu- and O-terminated $Cu_2O(111)$ surface planes). However, for Cu oxide surfaces the adsorption energy of the subsequently produced CO is significantly larger than on Cu metal, which decreases the predicted overall rate [13].

The present results confirm the CO formation rate is significantly increased on surfaces that have been covered with Cu_2O. However, we also show the enhancement persists well beyond the point at which most of the oxide layer has been reduced. While it is possible that some residual surface oxide species remain on the electrode following the main oxide reduction process, the lack of supporting theoretical results for CO_2 reduction on oxide surfaces noted above suggests this is not a promising model. Instead, we propose that the oxide reduction process leads to the formation of highly dispersed Cu nanoclusters which contain a large fraction of low-coordination number Cu surface atoms. The formation energy for the $COOH_{(a)}$ intermediate is likely to be lower for these sites, as indicated by the theoretical calculations for high-index Cu metal surfaces. (Recent calculations by others suggest that higher-index Cu surfaces will have different selectivities for higher hydrocarbon products, as well [14]).

While additional computational and experimental characterization work is needed to confirm this model, the present results suggest that efforts to develop better Cu-based electrocatalysts for CO_2 reduction should be focused on stabilizing highly dispersed Cu nanoparticles, as opposed to efforts at stabilizing the Cu oxide phase.

ACKNOWLEDGMENTS

This material is based upon work supported as part of the Center for Atomic Level Catalyst Design, an Energy Frontier Research Center funded by the U.S. Department of Energy, Office of Science, Office of Basic Energy Sciences under Award Number DE-SC0001058.

REFERENCES

1. A.T. Bell, in “Basic Research Needs: Catalysis for Energy”; U.S. Department of Energy, Office of Basic Energy Sciences Workshop August 6-8, 2007; available at: http://science.energy.gov/bes/news-and-resources/reports/basic-research-needs/
2. R. A. Kerr, *Science* **331**, 1510 (2011).
3. O. K. Varghese, M. Paulose, T. J. LaTempa, and C. A. Grimes, *Nano Letters* **9**(2) 731 (2009).
4. Y. Pan, et al., *Science* **333**, 988 (2011).
5. Y. Hori, “Electrochemical CO_2 Reduction on Metal Electrodes”, in *Modern Aspects of Electrochemistry*, **42**, edited by C. G. Vayenas, (Springer, 2008) p89-189.
6. M. Gattrell, N. Gupta, and A. Co, *Journal of Electroanalytical Chemistry* **594**, 1 (2006).
7. Karl W. Frese, Jr., *J. Electrochem. Soc.*, **138** (11) 3338 (1991).
8. T.-Y. Chang, R.-M. Liang, P.-W. Wu, J.-Y. Chen, and Y.-C. Hsieh, *Materials Letters* **63**, 1001 (2009).
9. M. Le, M. Ren, Z. Zhang, P. T. Sprunger, R. L. Kurtz, and J. C. Flake, *J. Electrochem. Soc.*, **158** (5), E45 (2011).
10. Y. Hori, A. Murata, and R. Takahashi, *JCS Faraday Trans. I*, **85** 2309 (1989).
11. D. T. Whipple and P. J. A. Kenis, *J. Phys. Chem. Lett.* **1** 3451 (2010).
12. A. Peterson, F. Abild-Pederson, F. Studt, J. Rossmeisl, and J. K. Norskov, *Energy & Environ. Sci.* **3** 1311 (2010).
13. A. Asthagiri (private communication)
14. W. Tang, A. A. Peterson, A. S. Varela, Z. P. Jovanov, L. Bech, W. J. Durand, S. Dahl, J. K. Nørskov and I. Chorkendorff, *Phys. Chem. Chem. Phys.*, **14** 76 (2012).

Mater. Res. Soc. Symp. Proc. Vol. 1446 © 2012 Materials Research Society
DOI: 10.1557/opl.2012.956

Porous Metal Oxides as Catalysts

Boxun Hu[1], Christopher Brooks[2], Eric Kreidler[2], Steven L. Suib[1, 3]*

1: Institute of Materials Science, University of Connecticut, Storrs, CT 06269, USA.

2: Honda Research Institute USA, Inc. 1381 Kinnear Rd. Suite 116, Columbus, OH 43212, USA.

3: Department of Chemistry, University of Connecticut, Storrs, CT 06269, USA.

ABSTRACT

Porous CoO/Mn_2O_3 Fischer-Tropsch (F-T) catalysts have been studied in CO hydrogenation. These CoO/Mn_2O_3 catalysts have been synthesized by incipient wetness impregnation method. These mesoporous catalysts have pore diameters of 2-25 nm and a surface area of 9.0 m^2/g. The gas and liquid products have been analyzed by an online gas chromatograph. The solid products were characterized by gas chromatograph-mass spectroscopy. These microsize cobalt catalysts exhibit good activity with 72.1% CO conversion and they are very stable in a 48 h stream test at 280°C. The selectivity to paraffins is above 95%. Few wax products were synthesized with a yield of less than 2%. The size effects of the cobalt catalysts have been studied by scanning electron microscopy.

INTRODUCTION

Fischer-Tropsch (F-T) synthesis has provided pathways of syngas (CO/H_2) to liquid fuel and useful chemicals. Existing catalysts (Co, Fe, Ru, and Ni) supported on inert silica, alumina, zeolites, and carbon nanotubes have been developed for F-T synthesis.[1-3] Cobalt and iron F-T catalysts have been used for industrial F-T synthesis.[4, 5] The products of these catalysts are mainly paraffins. Additional manganese oxides have been used as promoters to increase the production of olefins.[6, 7] The particle size affects the catalyst stability and activity.[8] Nanosize catalysts are favored due to their high surface area but sintering is a serious problem for using nanosize cobalt catalysts. In this study, microsize Mn_2O_3 has been selected as a new support for the study of the size effect.

These catalysts only contain cobalt, manganese, and oxygen. The simple components make these catalysts as a standard for the comparison with other existing promoted catalysts (e.g. K). For these purposes, the product profiles of gas, liquid, and solid products have been analyzed. The catalyst structure has been characterized by X-ray diffraction (XRD) and scanning electron microscopy (SEM).

EXPERIMENTAL

Catalyst Preparation

Materials used in this study were purchased from Alfa Aesar if no specification and they were used as received without further purification. The Mn_2O_3 supported cobalt catalysts were synthesized by an incipient wetness impregnation (IWI) method. For a typical synthesis, $Co(NO_3)_2$ (98+ wt%, molar ratio Co/Mn: 0.1-0.3) were dissolved in deionized (DI) water, and then Mn_2O_3 (98+ wt%, bixbyite-C) was added into the solution under agitation at 80°C until water was evaporated. The impregnated samples were dried at 120°C, and they were finally calcined in air (120 sccm) at 350-450°C in a tube furnace.

Characterizations of Catalysts and F-T Products

Powder XRD experiments were performed with a Scintag XDS 2000 X-ray diffractometer. SEM images were obtained using a Zeiss DSM 982 Gemini FESEM instrument. A Hewlett-Packard gas chromatograph (HP5890 series II) was used for gas chromatograph-mass spectroscopy (GC-MS) tests. Gaseous products were analyzed using an online GC (SRI 8610C). The components were analyzed by atomic adsorption analysis. . A Micromeritics ASAP 2020 instrument was used to measure Brunauer-Emmett-Teller (BET) surface areas and pore size distributions using nitrogen adsorption at -195°C.

Fischer-Tropsch Synthesis

Catalysts (0.8 g) were loaded into a packed stainless steel tube reactor (Inner diameter, 0.9 cm). The catalysts were reduced in 10% H_2/He, and then in 10% CO /He at a pressure of 1 atm and temperatures of 350-450°C for 2 h. CO Hydrogenation (H_2/CO: 2:1) has been tested in a packed bed reactor at temperatures of 120-320°C, a pressure of 13.6 atm, and a gas flow rate of 60 standard cm^3/min.

RESULTS AND DISCUSSION

During the calcination in air, $Co(NO_3)_2$ was converted to Co_3O_4 (JCPDS 9-418) and deposited on Mn_2O_3 supports. After reduced in H_2, Co_3O_4 converted to CoO (JCPDS 9-402) and face centered cubic Co (JCPDS 15-806) (Figure 1). Simultaneously, Mn_2O_3 supports changed to MnO (JCPDS 7-230). The XRD experiment was performed in air. Cobalt nanoparticles are readily oxidized to CoO and Co_3O_4 in air. When reduced in 10% CO/He, cobalt carbide phases have been reported in the catalysts.[9] For manganese oxide supported catalysts, an increased temperature and time are needed for a full reduction of cobalt catalysts. A high temperature of 450°C was used for the reduction compared to a reported temperature of 350°C for Co/Al_2O_3 catalysts. A strong metal-support effect hinders a full reduction.

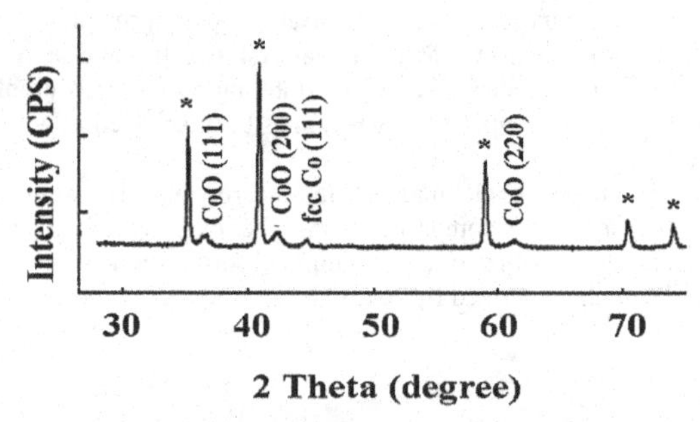

Figure 1. The XRD patterns of the reduced (H_2) Mn_2O_3 supported Co catalysts. "*" represents MnO.

In the IWI recipe, about 20 wt% of Co existed in the catalysts. The SEM images of Figure 2 show the morphologies and sizes of the catalysts. The sizes of the manganese oxide supports are 2-10 μ m (Figure 2A). Cobalt catalysts were coated on the manganese oxide supports, and the cobalt particle sizes are in the range of 30-100 nm (Figure 2B). The concentration of cobalt precursors, support sizes, and the calcination process affect the sizes of the cobalt catalyst. The average size of cobalt catalysts can be changed solely changing the cobalt precursor concentration (or Co/Mn ratio). The changes of product distributions of these FTS catalysts are being studied.

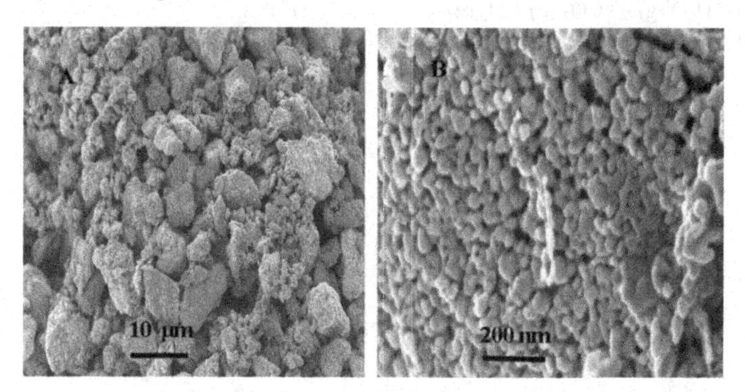

Figure 2. The SEM images of the reduced (H_2) Mn_2O_3 supported Co catalysts.

Figure 3 shows the pore size distribution of the Mn_2O_3 supported Co catalysts. The pore sizes of these catalysts are 4-25 nm. These microsize Mn_2O_3 catalysts have a BET surface area of 9.0 m^2/g. The BJH adsorption accumulative volume of pore between 17-5000 Å is 0.05 cm^2/g. The BJH desorption cumulative surface area of pore is 11.9 m^2/g.

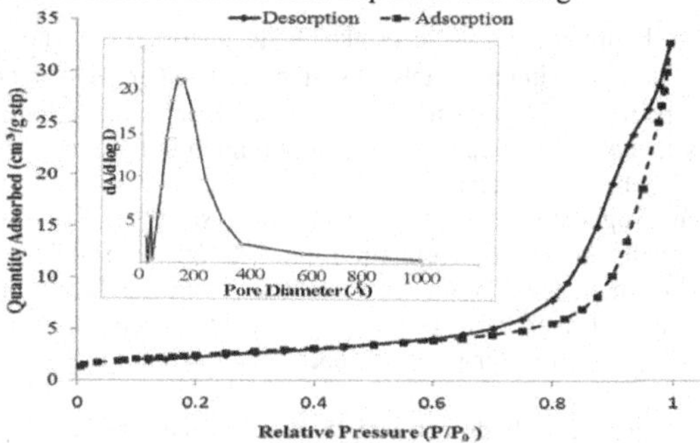

Figure 3. The BET surface area and pore size distribution of the post-FTS Mn_2O_3 supported Co catalysts.

The product distribution of hydrocarbons formed during the F-T synthesis follows an Anderson-Schulz-Flory (ASF) distribution, which can be expressed as:

$$Wn/n = (1-\alpha)^2\alpha^{n-1}$$

where Wn is the weight fraction of hydrocarbon molecules containing n carbon atoms. α is the chain growth probability. Figure 4 shows the ASF plot of the FTS products using the Mn_2O_3 supported Co catalysts. C1 (CH_4) formation does not follow the ASF distribution. Also, a large part (up to 27 wt%) of CO_2 has been produced via the water-gas shift reaction (Equation 1) and other reactions (e.g. Equation 2):

$$CO\ (g) + H_2O\ (g) \rightarrow CO_2\ (g) + H_2\ (g) \quad (1)$$

$$CO\ (g) + CO\ (g) \rightarrow CO_2\ (g) + C\ (s) \quad (2)$$

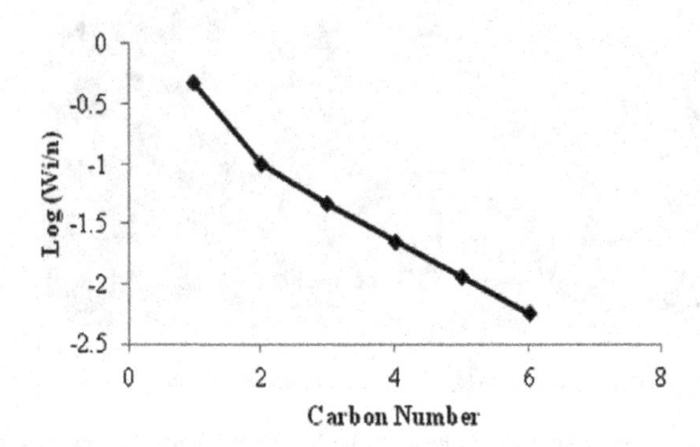

Figure 4. The ASF distribution of the FTS products using the Mn_2O_3 supported Co catalysts. The reaction temperature is 280°C.

The formation of C2-C6 hydrocarbons follows the ASF distribution. The chain growth probability α is 0.49. In this simple catalyst system, no other side reactions except dehydrogenation proceed in the F-T synthesis. Due to a thick layer of cobalt catalysts, the dehydrogenation effect by manganese oxide support is very limited in these catalysts. Therefore, a perfect ASF distribution has been demonstrated.

Due to lack of dehydrogenation effect as the above analysis, the F-T products are mostly paraffins, and only very small amounts of olefins present. The formation of large amounts of olefins has been reported in manganese promoted catalysts.[6, 7] Several factors may affect the ratios of olefin/paraffin in the F-T synthesis, such as H_2/CO, types of supports, and the nature of the catalysts. Due to high activity of water-gas shift reaction, a low ratio of H_2/CO will produce more olefins.

Most of products are light hydrocarbons, which is one of the characteristics of manganese oxide supported cobalt catalysts. Carboxylic acids and alcohol products were not detected by GC. A CO conversion of 72.1% is achieved under the above F-T conditions. These catalysts are very stable in a 48 h stream test. Very small amount of wax (<2 wt %) was produced. GC-MS data show that the carbon numbers of the wax products are 12 - 45, the most abundant is C21.

Table 1. The products distribution of the Mn_2O_3 supported Co catalysts

	C1	C2	C3	C4	C5	C6	>C7
Parafins (wt%)	45.12	18.43	12.53	8.19	5.15	3.09	4.75
Olefins (wt%)	-	0.97	0.66	0.43	0.27	0.16	0.25

CONCLUSIONS

In this study, porous CoO/Mn_2O_3 Fischer-Tropsch catalysts have been prepared and tested in CO hydrogenation. The particle sizes of cobalt oxide are about 30-100 nm in the CoO/Mn_2O_3 catalysts. These mesopore catalysts have a narrow distribution of 2-25 nm. These catalysts contained manganese promoters from the supports. High yields of CH_4 up to 45.12 wt% were produced. A CO conversion of 72.1% has been achieved using the cobalt catalysts (0.8 g) at a temperature of 280°C, a pressure of 13.6 atm, and a flow rate of 60 cm^3/min. The products are mainly C1-C6 paraffins, which follow the ASF distribution.

These Mn_2O_3 supported CoO catalysts have been used as reference catalysts due to their simple components and structure. The tests of size effects and stability have been performed using these catalysts. More complex F-T catalysts are being studied for selectivity synthesis of valuable products from the F-T synthesis.

ACKNOWLEDGMENTS

This project was funded by the Chemical Sciences, Geosciences, and Biosciences Division of the Office of Basic Energy Sciences, and Office of Science, U.S. Department of Energy and Honda Research Institute, USA. Inc.

REFERENCES

1. K. C. Mcmahon, S. L. Suib, B. G. Johnson, C. H. Bartholomew, *J. Catal.* **106**, 47-53 (1987).
2. S. L. Suib, K. C. Mcmahon, L. M. Tau, C. O. Bennett, *J. Catal.* **89**, 20-34 (1984).
3. N. Kruse, J. Schweicher, A. Bundhoo, A. Frennet, T. V. de Bocarme, *Top. Catal.* 48, 145-152 (2008).
4. B. H. Davis, *Ind. Eng. Chem. Res.* **46**, 8938-8945 (2007).
5. F. Fischer, H. Tropsch, *Brennst. Chem.* **7**, 97-116 (1926).
6. M. J. Keyser, R. C. Everson, R. L. Espinoza, *Appl. Catal.* A **171**, 99-107, (1998).
7. M. Ojeda, M. L. Granados, S. Rojas, P. Terreros, F. J. Garcia-Garcia, L.G. Fierro, *Appl. Catal.* A **261**, 47-55 (2004).
8. H. Karaca, J. P. Hong, P. Fongarland, P. Roussel, A. Griboval-Constant, M. Lacroix, K. Hortmann, O. V. Safonova, A. Y. Khodakov, *Chem. Commun.* **46**, 788-790 (2010).
9. J. P. den Breejen, P. B. Radstake, G. L. Bezemer, J. H. Bitter, V. Froseth, A. Holmen, K. P. de Jong, *J. Am. Chem. Soc.* **131**, 7197-7203 (2009).

Mater. Res. Soc. Symp. Proc. Vol. 1446 © 2012 Materials Research Society
DOI: 10.1557/opl.2012.1222

Role of Pt Nanoparticles in Photoreactions on TiO_2 Photoelectrodes

Woo-Jin An[1], Wei-Ning Wang[1], Balavinayagam Ramalingam[2], Somik Mukherjee[2], Dariusz M. Niedzwiedzki[3], Shubhra Gangopadhyay[2], and Pratim Biswas[1]
[1]Aerosol and Air Quality Research Laboratory, Department of Energy, Environmental and Chemical Engineering, Washington University in St. Louis, One Brookings Drive, Campus Box 1180, St. Louis, MO 63130, U.S.A.
[2]Center for Micro/Nano Systems & Nanotechnology, University of Missouri, Columbia, MO 65211, U.S.A.
[3]Photosynthetic Antenna Research Center (PARC), Washington University in St. Louis, One Brookings Drive, St. Louis, MO 63130, U.S.A.

ABSTRACT

Highly efficient Pt-TiO_2 composite photoelectrodes were synthesized by combining two novel deposition methods: ACVD and a room temperature RF (radio frequency) magnetron sputtering method. A room temperature RF magnetron sputtering method allowed uniform deposition of Pt nanoparticles (NPs) onto the as-synthesized nanostructured columnar TiO_2 films by ACVD. Pt NP sizes from 0.5 to 3 nm demonstrating a high particle density ($>10^{12}$ cm^{-2}) could be achieved by varying deposition time with constant pressure and power intensity. As-synthesized Pt-TiO_2 films were used as photoanodes for water photolysis. Pt nanoparticles deposited onto the TiO_2 film for 20s produced the highest photocurrent (7.92 mA/cm^2 to 9.49 mA/cm^2) and maximized the energy conversion efficiency (16.2 % to 21.2 %) under UV illumination. However, as the size of Pt particles increased, more trapping sites for photogenerated electron-hole pairs decreased photoreaction.

INTRODUCTION

Ever since Fujishima and Honda developed an innovative method to produce hydrogen (H_2) by water photolysis [1], much effort has been made to increase the energy conversion efficiency. Titanium dioxide (TiO_2) has been widely used as a photocatalytic material for solar energy applications. Along with the wide bandgap of TiO_2, its short electron-hole pair lifetime [2] is a limiting factor for decomposing water into oxygen (O_2) and hydrogen (H_2). One dimensional single crystal TiO_2 films provide favorable electron transport pathways, resulting in enhanced photoreaction [3]. In addition to the film morphology, surface modification by noble metal contacts lead to efficient electron-hole separation, improving photoelectrochemical properties of metal oxide films [4]. Although the size of noble metal particles and the distance between noble metal particles play a significant role in determining the efficiency of solar energy applications [5], existing methods can not precisely control these two factors. In this study, the room temperature RF magnetron sputtering method was employed to deposit monodispersed nano-sized platinum (Pt) metal on the columnar TiO_2 films in a controlled manner.

EXPERIMENT

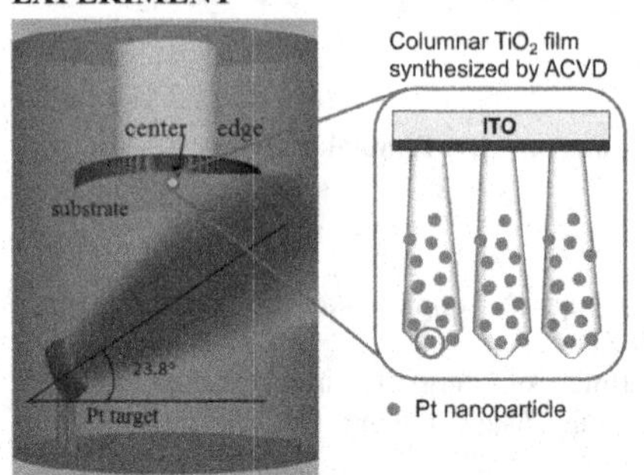

Figure 1. Sub-nanometer Pt particle deposition on the columnar TiO_2 film synthesized by ACVD [6].

Pt-TiO_2 photoelectrode synthesis was carried out in two steps. First, using aerosol chemical vapor deposition (ACVD), columnar TiO_2 films were deposited onto the ITO substrate (Delta Technologies, Stillwater MN) [7]. The thickness of each column was approximately 1.5 μm for a 50 min deposition by ACVD with 1.53 μmol/min of feedrate of titanium tetraisopropoxide (TTIP, Sigma-Aldrich, St. Louis MO) and a reactor residence time of 20 ms. The ITO substrate was maintained at 500°C. As shown in Figure 1, Pt NPs were deposited onto the as-synthesized nanostructured TiO_2 films by a room temperature RF magnetron sputtering with a tilting target capability (ATC 2000V, AJA International, Inc.) [8]. The sputtering conditions included a target angle of 23.8°, target-substrate distance of 6 inch, operating pressure of 4 mTorr, and fixed deposition power of 27 W. The sputtering time varied from 0s, 20s, 45s, and 200 s. The size of Pt NPs on the TiO_2 surface was characterized by high-resolution transmission electron microscopy (HR-TEM). The photocurrent density of Pt-TiO_2 photoelectrodes was measured with a three-electrode cell configuration. A Pt wire and an Ag/AgCl were used as a counter electrode and a reference electrode respectively. 1M of KOH was used as an electrolyte, which pH was 14. A potentiostat (BASi Inc. model CV-50W, West Lafayette, IN) was to scan the voltage with 10 mV/s. A Xenon arc lamp (Oriel model 66021, Stratford, CT) was used as light source, which light intensity was 40 mW/cm^2 in the UV regime. The time-resolved transient absorption (TA) spectroscopy measurements were carried out using a femtosecond TA spectrometer (Helios, Ultrafast Systems, LCC). The samples were excited at 365 nm with pump beam energy kept at low energies between 1 and 3 μJ. The excitation beam was focused to a spot size of 1 mm diameter corresponding to photon intensity of $2.34 \sim 7.04 \times 10^{14}$ $photons/cm^2$.

DISCUSSION

The size and interparticle distance of Pt NPs are the determinant parameters of overall photoenergy conversion efficiency. Under UV illumination, in the pristine TiO_2 film, the photogenerated electrons can typically either recombine with holes or transfer to the counter electrode. However, when Pt NPs form contact with the TiO_2 surface, electrons can be stored in

Pt NPs, reacting with O_2 a resulting in superoxide radical formation ($O_2^{\bullet-}$) [5]. The work function of Pt NPs varies with respect to their sizes. Electron transfer from TiO_2 to Pt NPs, occurs only when the work function of Pt NPs is lower than the conduction band of TiO_2 (-4.4 eV in vacuum). If the size of Pt NPs is too small, the transfer of electrons to the TiO_2 is inhibited by the higher work function of Pt relative to the conduction band of TiO_2. If its size is too large, Pt NPs can be oxidized by photogenerated holes. In addition to the size effects, the interparticle distance between Pt NPs can also influence the behavior of photogenerated electrons and holes.

Growth of Pt nanoparticle deposited on the TiO_2 surface

Table 1 describes the size change of Pt NPs on the TiO_2 surface with different RF sputtering times: 20s, 45s, and 200s. The unique tilt target configuration aids in using the low density, low energy metal atom of the deposition flux, in the nanoparticle growth. As the deposition time increases Pt nanoparticles nucleate and grow in defect sites, following a Volmer-Weber growth mode. After reaching a size-limiting factor, they start to coalesce forming larger spherical Pt nanoparticles with narrower interparticle distance, with the use of these low energy atoms of the deposition flux, we gain excellent control over the size and density of the Pt nanoparticle formation.

Table 1. Summary of mean diameters and standard errors from Gaussian distribution of Pt NPs as a function of sputtering time (20s, 45s, and 200s) [6].

Pt sputtering time [s]	Mean diameter of Pt nanoparticle [nm]	Std error
20	1.15	0.005
45	1.34	0.010
200	3.46	0.072

Characteristics of TiO_2 photoelectrodes with different sizes of Pt nanoparticles

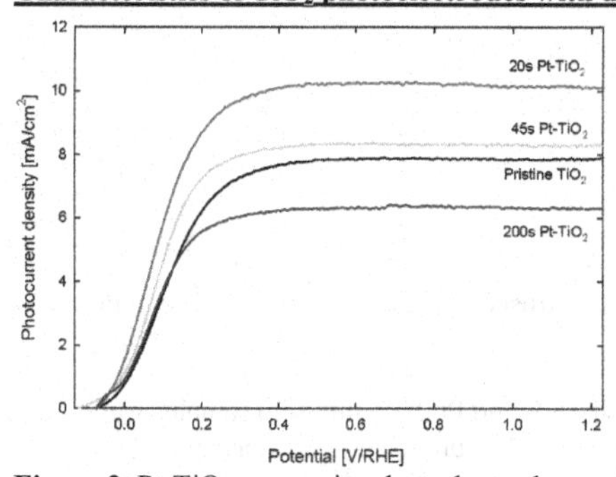

Figure 2. Pt-TiO_2 composite photoelectrodes performance characteristics: photocurrent as a function of bias potential [6].

Figure 2 presents IV characteristics of as-synthesized Pt-TiO_2 photoelectrodes and corresponding results are summarized in Table 2. Compared to the pristine TiO_2 photoelectrode, 1.15 nm Pt NPs from 20 s of RF sputtering contributed to both the increased photocurrent density (7.92 mA/cm^2 to 9.49 mA/cm^2) and the higher energy conversion efficiency (16.2 % to 21.2 %). The effective separation of electron-hole pairs followed by the reaction on the Pt NPs surface was advantageous. For 45 s deposition, the size of Pt NPs slightly increased to 1.34 nm. This Pt-TiO_2 photoelectrode showed still improved photocurrent density and efficiency than the pristine TiO_2 photoelectrode, but, less than the 20s sputtering of Pt- TiO_2 film. The work function was slightly shifted due to the size increase of Pt NPs. However, interparticle distance between Pt NPs became narrower, as more Pt NPs formed on the TiO_2 surface. Thus, photogenerated electrons and holes would have greater opportunity to recombine compared to the 20s deposited Pt-TiO_2 film. In the case of 200s deposition of Pt NPs, the size of Pt NPs is large enough to act as a recombination center.

Table 2. Summary of characteristics of as-synthesized Pt-TiO_2 composite photoelectrodes [6].

Pt sputtering time [s]	Saturated current [mA/cm^2]	Fill factor	Efficiency [%]
0	7.92	0.66	16.2
20	9.49	0.70	21.2
45	8.29	0.68	17.7
200	6.31	0.70	13.8

Transient absorption of Pt-TiO_2 photoelectrode

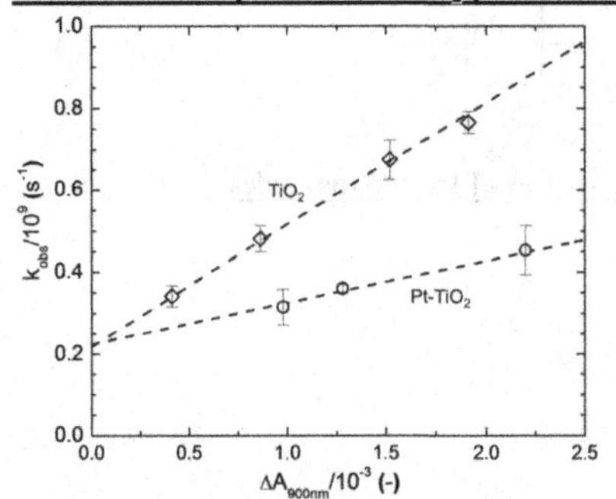

Figure 3. Time-resolved transient absorption (TA) spectroscopy measurement of TiO_2 and Pt-TiO_2 thin films (Pt loading time of 45 s) [9].

The charge carrier transfer dynamics for both TiO_2 and Pt-TiO_2 thin film samples were analyzed by femtosecond time-resolved TA spectroscopy measurements. Dependency of the electron-hole pair recombination rate constant on the pair concentration for both TiO_2 and Pt-TiO_2 films is shown in Figure 4. In both cases charge recombination rates are dependent on the excitation intensity (electron-hole pair concentration) and the observed trends can be fitted by linear functions. It is apparent that the linear fitting of the Pt-TiO_2 film has a lower slope compared to its TiO_2 counterpart, indicating that the charge recombination process is slower for Pt-TiO_2. This is the direct evidence of Pt-driven suppression of the charge recombination process in the Pt-TiO_2 sample, which is similar to that of Au-TiO_2 [10].

CONCLUSIONS

Two novel deposition methods, ACVD and a room temperature RF magnetron sputtering method, were used to synthesize the Pt-TiO_2 photoelectrodes. The size of Pt NPs deposited on the TiO_2 surface was precisely controlled with different sputtering times. Under UV illumination, the 20s deposited Pt-TiO_2 photoelectrode, where 1nm Pt NPs were sparsely deposited onto the TiO_2 surface, increased photocurrent density (7.92 mA/cm^2 to 9.49 mA/cm^2) and enhanced energy conversion efficiency (16.2 % to 21.2 %), compared to the pristine TiO_2 photoelectrode. However, the photoelectrochemical performance was diminished for both larger Pt NP sizes and smaller interparticle distance.

ACKNOWLEDGMENTS

This material is based upon work supported as part of the Photosynthetic Antenna Research Center (PARC), an Energy Frontier Research Center funded by the U.S. Department of Energy, Office of Science, Office of Basic Energy Sciences under Award Number DE-SC 0001035. Partial support from the NNIN NRF at Washington University is also acknowledged.

REFERENCES

1. A. Fujishima, K. Honda, *Nature* **238**, 37-38 (1972).
2. G. Rothenberger, J. Moser, M. Grätzel, N. Serpone, and D. K. Sharma, *J. Am. Chem. Soc.* **107**, 8054-8059 (1985).
3. R. van de Krol, Y. Liang, and J. Schoonman, *J. Mater. Chem.* **18**, 2311-2320 (2008).
4. A. Hagfeldt and M. Grätzel, *Chem. Rev.* **95**, 49-68 (1995).
5. M. Sadeghi, W. Liu, T.-G. Zhang, P. Stavropoulos, and B. Levy, *J. Phys. Chem.* **100**, 19466-19474 (1996).
6. W.-J. An, W.-N, Wang, B. Ramalingam, S. Mukherhee, B. Daubayev, S. Gangopadhyay, and P. Biswas, *Langmuir* **28**, 7528-7534 (2012).
7. W.-J. An, E. Thimsen, and P. Biswas, *J. Phys. Chem. Lett.* **1**, 249-253 (2010).
8. M. Yun, B. Ramalingam, and S. Gangopadhyay, *Nanotechnology* **22**, 465201 (2011).
9. W.-N. Wang, W.-J. An, B. Ramalingam, S. Mukherjee, D. M. Niedzwiedzki, S. Gangopadhyay, P. Biswas, *Submitted* (2012).
10. V. Subramanian, E. E. Wolf and P. V. Kamat, *J. Am. Chem. Soc.* **126**, 4943-4950 (2004).

Mater. Res. Soc. Symp. Proc. Vol. 1446 © 2012 Materials Research Society
DOI: 10.1557/opl.2012.957

Metal Oxides as Catalyst Promoters for Methanol Oxidation

Praveen Kolla[1], Kimberly Kerce[2], Hao Fong[2] and Alevtina Smirnova[2]
[1]Material Engineering Science Program and [2]Department of Chemistry, South Dakota School of Mines and Technology, Rapid City, SD 57701, U.S.A.

ABSTRACT

The Transition metal oxides such as TiO_2 and CeO_2 as catalyst and co-catalyst materials were studied for methanol oxidation. The metal oxide nanoparticles were impregnated into carbon aerogel and Pt-Ru/C (Tanaka) by modified sol-gel Pechini method and heat-treated at different temperatures. Crystal structure, particle size and composition of the catalyst particles were studied using XRD, TEM and EDS techniques. The electrochemical activity and stability of these catalyst materials were studied in acidic medium and the results were compared to their corresponding specific and active surface areas. The aerogel supported metal oxides were stable and proved for better methanol oxidation, while a significant synergetic effect in electro-oxidation is observed when the metal oxides were impregnated into the structure of Pt-Ru/C catalyst. The methanol-oxidation was further improved after heat treatment due to its improved structural and surface properties.

INTRODUCTION

Direct Methanol Fuel Cells (DMFC) is a PEMFC system, which uses renewable methanol as fuel. The high specific-energy density of methanol fuel will greatly simplifies the storage complications associated with hydrogen/air systems and mobility problems with rechargeable battery based portable electronic devices [1]. However, slower anode kinetics, CO-poisoning, chemical instability of catalyst in acidic environment and methanol crossover to cathode side are some of the challenges to DMFC commercial viability. At this point, cost reduction through exploration of noble and non-noble metal based anode catalysts, development of more stable and conductive carbon supports, and alloying approach are the current R&D strategies [2].

Some metal oxides are known to show electrochemical catalytic activity because of metal ability to switch between different valences. These metal oxides are also known to exhibit metal-like conductivity when partially filled d- and f-bands are available [3]. While Pt-Ru based materials are well known as anode catalysts for methanol oxidation, recent efforts have been focus to promote these catalysts with non-noble metal oxides for better stability and for enhanced oxidation of methanol and CO [4-7].

Advancements in the nanomaterials lead to the development of more stable and high-surface area hybrid-carbon fuel cell supports. Carbon aerogels supports are commercially feasible materials for its scalability of its fabrication and simple catalyst impregnation methods [8]. However, the efficient utilization of catalyst is possible through optimization of these materials in terms catalyst dispersion, electrochemical surface area and pore size distribution for different impregnation methods [2]. In this regard, carbon aerogel supported transition metal oxide catalysts and metal oxide impregnated Pt-Ru/C commercial catalyst was studied for methanol oxidation.

EXPERIMENT

Catalyst Synthesis

A modified sol-gel (Pechini) combustion synthesis [9] is used to impregnate metal oxides on carbon aerogel (aspen aerogels®) and Pt-Ru supported carbon (Tanaka®). Cerium nitrate hexahydrate ($Ce(NO_3)_3.6H_2O$) and titanium (IV) isopropoxide ($Ti\{OCH(CH_3)_2\}_4$) are used as precursors. In this method, initial mixture is prepared by dissolving metal organic precursor and glycine in the equal quantities of isopropanol and water. In case of TiO_2 synthesis, 1mL of concentrated HNO_3 is added to dissolve precursor in the glycine mixture and to form glycine-nitrate complex. This viscous mixture is added to the supporting material for impregnation of metal oxide. A porous mixture is formed after drying it in the air at 50°C. The synthesized material is gradually heated to 350°C for glycine-nitrate combustion, which uniformly disperses metal oxide nanoparticles on the support. The resulting catalyst powder is further heated at 600°C for 1 hour in inert gas atmosphere (N_2 gas) to develop poly-crystalline nanocomposites.

Catalyst Characterization

The crystal structures of CeO_2 and TiO_2 on carbon aerogel or Pt/Ru-C, heated at 350°C and at 600°C are characterized by Rigaku Ultima Plus theta-theta X-ray diffractometer. Cu-kα radiation (λ = 1.54178 Å) is used to scan the materials from 10°-90° (2θ) with scan rate of 0.667°/min. The particle morphology is studied using high resolution JEOL JEM-2100 LaB_6 transmission electron microscope at 200 keV. The catalyst powders are dispersed in ethanol by sonication and placed on square mesh copper grid coated with carbon film (CF200-Cu) for analysis. The compositions of metal oxides are estimated with an Oxford Inca energy-dispersive silicon-drift X-ray (EDX) spectrometer.

The electrochemical activity of the catalysts is studied in conventional three electrode half-cell using 0.1M $HClO_4$ as electrolyte with reference to Ag/AgCl/0.1M KCl. The catalyst inks for Rotating Disc Electrode (RDE) measurements was prepared by mixing 0.4 mL of 5% 1100 Nafion™ solution in 20 mL of isopropanol and 79.6 mL of water to 0.6g of catalyst powder. The resulting catalyst ink is sonicated to disperse the catalyst particles and 10μL of the ink is dried to form a thin film on glassy carbon electrode (having 0.19 cm^2 surface area). The electrolyte is purged with nitrogen for 30 minutes prior to the experiment. The RDE-Cyclic Voltammetry (RDE-CV) is studied using a Pine instruments setup at 1600 RPM by adding 2M methanol to electrolyte while keeping a blanket of nitrogen flowing on the electrolyte surface. Same setup is also used to study stability of the catalysts with chronoamperometry experiment.

Chemisorb 2720 from Micromeritics® is used for the BET specific surface area and active surface area analysis. The BET specific surface area is measured by N_2 adsorption-desorption isotherms at 77K by degassing the catalyst powders at 200°C for 30 minutes. Pulsed-chemisorption technique is used to find active surface area of the catalysts. The catalyst powders are degassed at 350°C for 1 hour by passing Argon gas and cooled to room temperature prior to the analysis. In the analysis, 1mL pulses of 10% active H_2 in Argon gas are injected by flowing Argon (20mL/min) through the sample and the Thermal Conductivity Detector (TCD) signal is recorded till the surface area of the pulse get stabilized. The total amounts of the hydrogen consumed per unit gram of the samples are calculated from the surface area data and the corresponding active surface areas of the catalysts are calculated.

DISCUSSION

The X-ray diffraction intensities of heat treated CeO_2, and TiO_2 nanoparticle impregnated samples are presented in the Figure 1(a) & 1(b) respectively. The samples heated at 600°C shows high crystalline structure. The carbon aerogel supported CeO_2 shows peaks at (2θ values) of 28.7°, 47.7° and 56.6° and CeO_2 impregnated on PtRu/C have peaks corresponding to (2θ values) of 28.7°, 33.2°, 47.7°and 69.8° (Figure 1(a)). These diffraction planes of crystalized CeO_2 corresponding to cubic crystal structure of the fluoride type oxide [10]. The peak at 43.5° corresponding to Ce-Ru alloy is evidence of possible electronic interaction between cerium and ruthenium. The 600°C heated TiO_2 samples in Figure 1(b) shows the presence anatase-TiO_2 crystalline structure along with the rutile form.

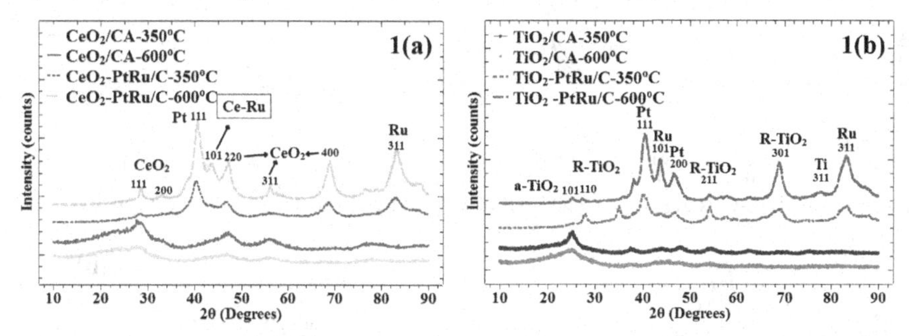

Figure 1: The comparison of X-ray diffraction analysis of (a) CeO_2 and (b) TiO_2 samples

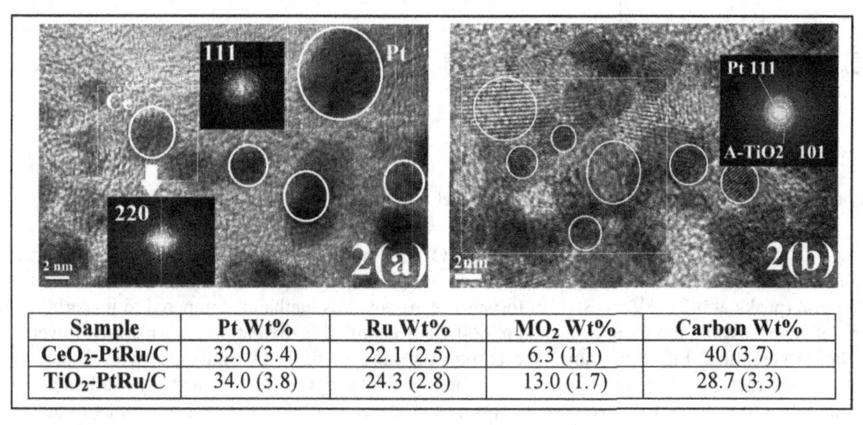

Sample	Pt Wt%	Ru Wt%	MO_2 Wt%	Carbon Wt%
CeO_2-PtRu/C	32.0 (3.4)	22.1 (2.5)	6.3 (1.1)	40 (3.7)
TiO_2-PtRu/C	34.0 (3.8)	24.3 (2.8)	13.0 (1.7)	28.7 (3.3)

Figure 2: TEM and EDX analysis of (a) CeO_2-PtRu/C and (b) TiO_2-PtRu/C

The TEM images (In the Figure 2) show the CeO_2 and TiO_2 particles are well-dispersed with Pt-Ru nanoparticles and the size of these particles is about 2-5 nm. The lattice parameters from FFT-TEM images show the presence of CeO_2 [220] (lattice parameter of 0.19 nm), Pt [111]

(lattice parameter of 0.22 nm) and anatase-TiO_2 (lattice parameter of 0.35 nm). The formation of uniform and highly dispersed nanoparticles of metal-oxides is due to the uniform mixing of the precursor at atomic and molecular levels and rapid combustion [11]. The EDS compositional analyses of the particles are tabulated in the Figure2. The weight percent of Pt, Ru, carbon and metal oxides are presented in the table with errors in brackets.

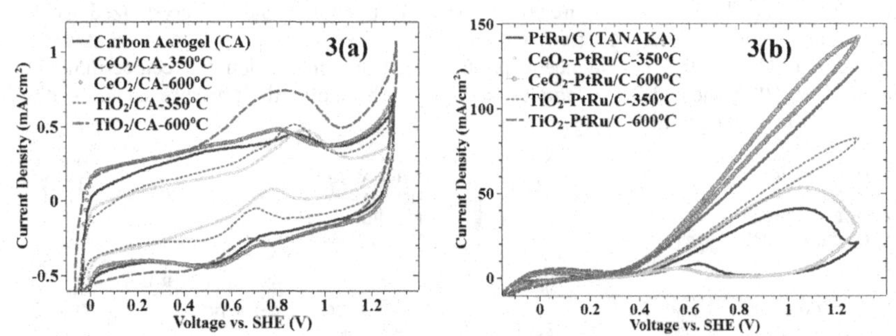

Figure 3: The RDE-CVs of (a) CeO_2 and TiO_2 on CA (b) CeO_2 and TiO_2 on PtRu/C (Tanaka)

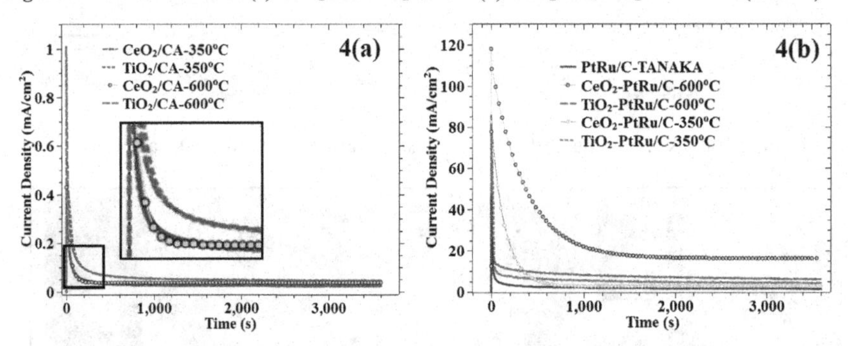

Figure 4: Chronoamperometry of (a) TiO_2 and CeO_2on CA (b) TiO_2 and CeO_2 on Pt-Ru/C

The normalized current densities of (a) TiO_2/CA and CeO_2/CA and (b) TiO_2-Pt-Ru/C and CeO_2-Pt-Ru/C catalyst are presented in the Figure 3. In Fig 3(a), the metal oxides shows oxidation (peaks at 0.7-0.9V vs. SHE in forward scan), towards methanol compared to its carbon aerogel support while the samples heated at 600°C have shifted its oxidation peak towards lower oxidation potentials. Fig. 3(b) shows the current densities of the TiO_2 and CeO_2 with Tanaka catalyst. The metal oxide infused into PtRu/C catalyst shows higher catalytic activity (both at 0V and at 1.1V) than the PtRu/C (blue-line), primarily heated at 600°C shows 2.5-3 times higher activity than PtRu/C catalyst. The high oxidation peak to reduction peak is observed for CeO_2 heated at 350°C while the oxidation peaks of the other samples shifted to higher voltages. The impregnated metal oxides have shown higher methanol oxidation with the Pt-Ru/C than its independent activity as metal oxides could electronically interact and participate in CO-oxidation along with ruthenium [12, 13]. The further understanding of the reaction mechanism could help in understanding this electronic interaction.

The comparative stability of the metal oxide on CA catalysts at its oxidation peaks (0.7-0.9 V) are presented in the Fig 4(a), whereas the stability of metal oxide impregnated PtRu/C catalyst at 1.1V are compared in the Fig. 4(b). From Fig. 4, it is apparent that TiO_2/CA heated at 600°C and CeO_2 incorporated PtRu/C heated at 600°C shows better stability.

Sample	BET-Surface Area (SSA) m^2/g (error)	Volume H_2-Chemisorbed µ-mole/g of catalyst (error)
Carbon Aerogel (CA)	860 (23)	-
TiO_2/CA-350°C	663 (20)	95.6 (4.3)
TiO_2/CA-600°C	691(22)	72.5 (3.7)
CeO_2/CA-350°C	520 (25)	112 (5.6)
CeO_2/CA-600°C	572 (30)	91 (4.5)
PtRu/C (Tanaka)®	220 (15)	265 (3.8)
TiO_2-PtRu/C-350°C	77 (9)	311 (10.2)
TiO_2/PtRu/C-600°C	152 (11)	277 (11.3)
CeO_2/PtRu/C-350°C	86 (14)	288 (9.8)
CeO_2/PtRu/C-600°C	170 (12)	272 (7.7)

Table 1: The BET- surface Area and Pulsed Chemisorption analysis data

The specific surface area (SSA) and the hydrogen pulsed-chemisorption data of the samples is provided in Table 1. The SSA of the metal oxide nanoparticles heated at 600°C slightly increased in comparison to the corresponding samples heated at 350°C. The room temperature pulse-chemisorption studies show that the samples heated at 600°C adsorb slightly lower volumes of H_2 gas compared to samples heated at 350°C. The lower BET specific surface area of the samples heated at 350°C compared to the ones heated at 600°C could be due to the blocking of microspores by unburned metal-organic precursors [14].

CONCLUSIONS

The nanoparticles of TiO_2 and CeO_2 (2-5 nm) were impregnated into the carbon aerogel support and Pt-Ru/C catalyst by a modified sol-gel Pechini method. According to the surface area studies the produced mixed ceramic-inorganic composites demonstrated high specific surface area and high catalytic activity. The metal oxides impregnated into the Pt-Ru/C exhibited higher methanol oxidation compared to the pure metal oxide phase. The XRD studies showed that the CeO_2 on PtRu/C catalyst heated at 600°C formed a Ce-Ru complex. The presence of platinum group metals would significantly lower the reduction temperatures of CeO_2. In this case, Ceria can be irreversibly reduced by heating at 600°C in nitrogen [15]. This will cause the enhanced electronic interaction between cerium and ruthenium improving oxidation of carbon monoxide during electrooxidation of methanol. On the contrary, TiO_2 sample heated at 600°C demonstrated the presence of TiO_2 on PtRu/C as a mixture of anatase and rutile phase. According to the previously published results, anatase phase can greatly improve the electro-oxidation of methanol [16]. The amperometric studies show that among titanium and cerium oxides, TiO_2/CA heated at 600°C has the highest stability in perchloric acid solutions. Regarding titanium and cerium oxides impregnated into the structure of PtRu/C, the stability of CeO_2-PtRu/C (600°C) is the highest due to the formation of Ce-Ru alloy.

ACKNOWLEDGMENTS

This work is supported by NSF/EPSCOR grant no: 0903804 and state of South Dakota. Authors would like to acknowledge Dr. Phil Ahrenkiel, Srujan Mishra and Dr. Rajesh Shende for the help provided while TEM analysis and heat treatment of the materials.

REFERENCES

1. S.K. Kamarudin, F. Achmad and W.R.W. Daud, International Journal of Hydrogen Energy, 34, 16 (2009).
2. Hansan Liu, Chaojie Song, Lei Zhang, Jiujun Zhang, Haijiang Wang, and David P. Wilkinson, Journal of Power Sources, 155 (2), 95-110 (2006).
3. B. Viswanathan, Ch. Venkateswara Rao and U. V. Varadaraju, Energy and Fuel, 43-101(2006).
4. Alexey Serov and Chan Kwak, Applied Catalysis B: Environmental, 90, 3–4 (313-320), 17 (2009).
5. Tao Huang, Deng Zhang, Leigang Xue, Wen-Bin Cai, and Aishui Yu, Journal of Power Sources, 192 (2), 285-290 (2009).
6. K Lasch, L Jörissen, and J Garche, Journal of Power Sources, 84 (2), 225-230 (1999).
7. Z Jusys, T.J Schmidt, L Dubau, K Lasch, L Jörissen, J Garche, and R.J Behm, Journal of Power Sources, 105 (2), 297-304 (2002).
8. C. Moreno-Castilla and F.J. Maldonado-Hódar, Carbon, 43 (3), 455-465 (2005).
9. Tianyou Peng; Xun Liu; Ke Dai; Jiangrong Xiao and Haibo Song, Materials Research Bulletin, 41, 1638–1645 (2006).
10. Abbasi, Z., Haghighi, M., Fatehifar, E. and Rahemi, N, Asia-Pacific Journal of Chem. Eng. (2011).
11. Zongping Shao, Wei Zhou and Zhonghua Zhu, Progress in Materials Science, 57 (4), 804-874 (2012).
12. M.W. Roberts and M. Tomellini, Catalysis Today, 12 (4), 443-452 (1992).
13. M. Aulice Scibioh, Soo-Kil Kim, Tae-Hoon Lim, Seong-Ahn Hong, and Heung Yong Ha, ECS Trans. 6, 93 (2007).
14. C. Moreno-Castilla and F.J. Maldonado-Hódar, Carbon, 43 (3), 455-465 (2005).
15. B. Harrison, A. F. Diwell and C. Hallett, Promoting Platinum Metals by Ceria, Platinum Metals Rev., 1988, 32, (2).
16. Roderick E. Fuentes, Brenda L. Garcı´a and John W. Weidner, Journal of Electrochemical Society, 158 (5) B461-B466 (2011).

Mater. Res. Soc. Symp. Proc. Vol. 1446 © 2012 Materials Research Society
DOI: 10.1557/opl.2012.989

Supported Ni catalyst made by electroless Ni-B plating for diesel autothermal reforming

Zetao Xia[2], Liang Hong*[1,2], Wei Wang[2], Zhaolin Liu[2]

[1] Department of Chemical and Biomolecular Engineering, National University of Singapore, Singapore, 117576

[2] Institute of Materials Research & Engineering, 3 Research Link, Singapore, 117602

ABSTRACT

$Ce_xGd_{1-x}O_{2-\delta}$ (CGO)-supported Ni nano grains were prepared initially by electrolessly depositing Ni-B alloy nano particles onto an activated carbon (AC). The as-deposited Ni-B particles were then transferred from AC to CGO through the metallo-organic precursor approach. The resultant Ni/CGO catalysts displayed excellent catalytic activity and chemical stability against coking and sulfur poisoning in catalyzing autothermal reforming (ATR) of a surrogate diesel fuel, comprising dodecane, tetralin and a substituted thiophene. For comparison purpose, a Ni/CGO catalyst prepared by the conventional impregnation method was employed in the same ATR system. These two catalytic systems exhibited rather discrepant outcomes. It was found that the Ni(B)/CGO catalyst was capable of repressing selectivity of ethylene during the reforming process. In addition to this, CGO played a critical role in thermal cracking hydrocarbon chains and inhibiting sulfur poisoning.

INTRODUCTION

The solid oxide fuel cell (SOFCs) and proton exchange membrane fuel cell (PEMFC) have higher energy efficiency than internal combustion engine and can be fuelled by synthesis gas or hydrogen generated from reforming infrastructure fuel, such as diesel, using as an auxiliary power unit (APU) in automotive applications [1-3]. The supported-Ni catalyst system is most desired for the reforming of a liquid fuel due to its low cost and high activity. However, the Ni-based catalyst usually cannot inhibit carbon deposition, or coking, and sulfur poisoning. Such deactivation will become more severe with the increase in complexity of hydrocarbon molecules. The use of an oxygen conducting solid electrolyte in particular ceria as support for Ni metal clusters [4-7] is an effective solution to the deactivation because ceria offers not only strong stability to sintering but also surface oxygen vacancies. In addition, a part of CeO_2 is reduced to Ce_2O_3 under reducing atmosphere. The Ce(III) species induces the sulfidation reaction and hence helps mitigate sulfur poisoning [8]. Such sacrificial role reduces sulfur poisoning at the Ni catalytic sites [9]. In the present study, a special chemical modification on the Ni atomic cluster was attempted. This method is illustrated in Figure 1 in which nanoparticles of Ni-B alloy were initially developed on an activated carbon (AC) by utilizing its high surface area and incinerable property. The AC-loaded Ni-B was uniformly dispersed in a metallo-organic gel of Ce^{3+}-Gd^{3+} ions. When the mixture was subjected to pyrolysis and heat treatment, a CGO-supported NiO-B_2O_3, the precursor of the catalyst, was obtained. This two-step synthesis process, termed as nanoparticle transfer, benefited the distribution of NiO in/on the CGO phase to a large extent because of the in-situ design and negligible chemical reactivity between NiO and CGO.

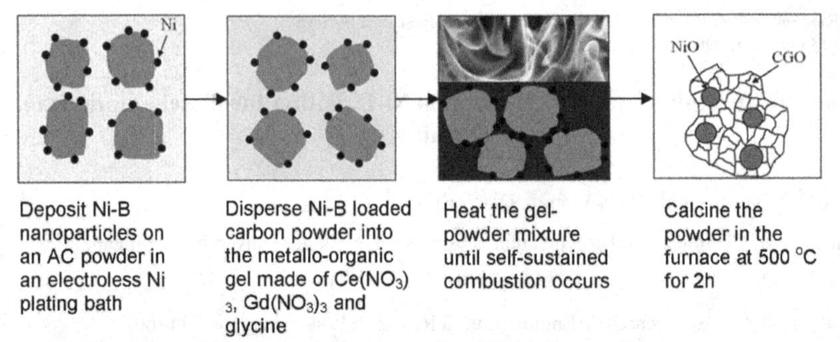

Figure 1. Flow chart of the preparation and transfer of Ni-B nanoparticles

EXPERIMENTAL DETAILS

The CGO ($Ce_{0.9}Gd_{0.1}O_{1.95}$)-supported Ni catalysts were synthesized by the following three procedures, respectively. *a.* By impregnation starting from immersing the CGO powder, prepared by incinerating the metallo-organic gel of Ce^{3+}-Gd^{3+} ions, in an aqueous solution of Ni nitrate salt; *b.* The nanoparticle transfer method by using hydrazine (N_2H_4) as reducing agent; *c.* using potassium borohydride (KBH_4) instead of hydrazine in the same procedure as *b*. An alloy composition of Ni_{17}-B was obtained from this metal deposition. Both ceria supported systems, *b* and *c*, were treated at 500 °C for 2 h before use.

These three prepared catalysts were assessed in a micro fixed-bed reactor operated under ATR condition. The reactor was loaded with about 0.21 g of the catalyst. A surrogate diesel fuel composed of dodecane (75 wt.%) tetralin (25 wt.%) and 3-methyl-benzothiophene (MBTP, 50 ppm) was designed for ATR. The temperatures of preheater at 250°C and of reactor at 750°C, the gas hourly space velocity of 5000 h^{-1}, the H_2O/C ratio of 3.0, and the O/C ratio of 0.7 were set to undertake ATR of the surrogate fuel. The product stream was arranged to pass a cold trap (solid ice) and then sent to an on-line gas chromatography (Shimadzu GC-2010) to determine the composition of reforming reaction.

DISCUSSION

Autothermal reforming of diesel surrogate: activity and deactivation

The three Ni/CGO catalysts described above were named as Ni(imp)/CGO for the one made by the impregnation method; Ni(H)/CGO by hydrazine-reduced electroless; and Ni(B)/CGO by boronhydride-reduced electroless. The Ni(imp)/CGO catalyst exhibited an overall decreasing conversion with reaction (Fig. 2). In addition, the slight decrease in H_2 selectivity and increase in CH_4 selectivity meant deactivation of the catalyst. On the contrast, both Ni(H)/CGO and Ni(B)/CGO catalysts displayed relatively more steady conversion and selectivity (Figs. 3 – 4) than Ni(Imp)/CGO. This suggests that the nanoparticle transfer method be more effective in dispersing Ni catalytic sites over the ceria support than the impregnation method. Then, compared with Ni(H)/CGO, Ni(B)/CGO exhibited better conversion despite similar average selectivity data of H_2, CO_2, CO and CH_4, which were about 61.4, 23.6, 10.9 and 3.7 %, respectively. It was also noticed that negligible B_2O_3 was detected by XPS and ICP in the Ni(B)/CGO after ART. Although how the B_2O_3 phase impacted the catalysis is still not clear, we are inclined to consider that it minimize the interfacial difference between Ni and CGO and be likely removed through reduction to borane during ATR. As a result, the proposed interfacial mediation role of B_2O_3 might lead to strong metal-support interactions (SMSI).

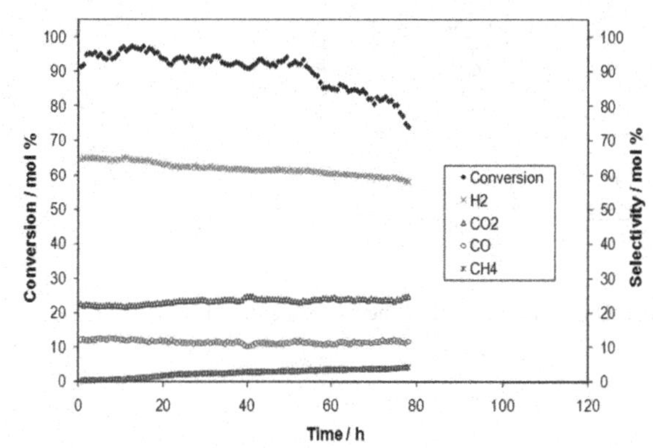

Figure 2. Assessment of the Ni(imp)/CGO catalyst in ATR of the surrogate diesel.

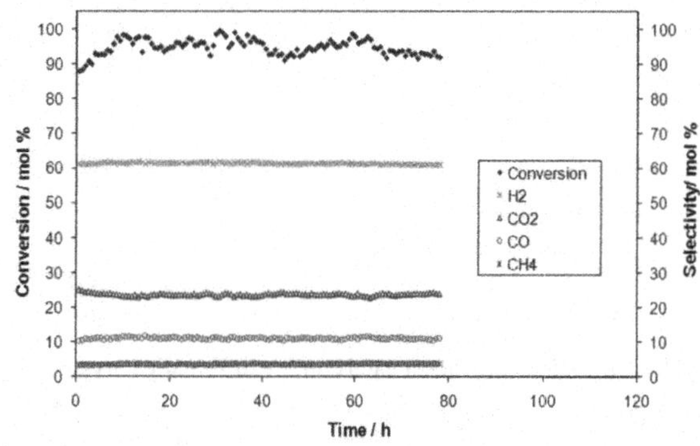

Figure 3. Assessment of the Ni(H)/CGO catalyst in ATR of the surrogate diesel.

Selevtivity of C_2H_4

Ethylene (C_2H_4) was reported to cause carbon formation in the reforming process [9]. As shown in Fig. 5, the selectivity of ethylene increased dramatically over the Ni(Imp)/CGO catalyst with the moving of reaction. But this trend became much alleviated over the Ni(H)/CGO catalyst and trivial over the Ni(B)/CGO catalyst in the ATR duration of 120 h.

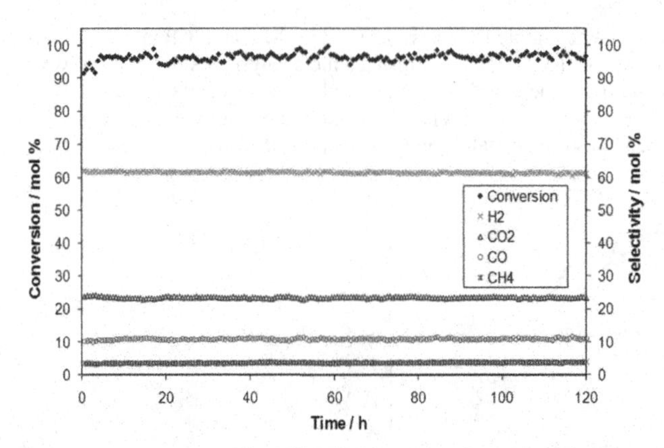

Figure 4. Assessment of the Ni(B)/CGO catalyst in ATR of the surrogate diesel.

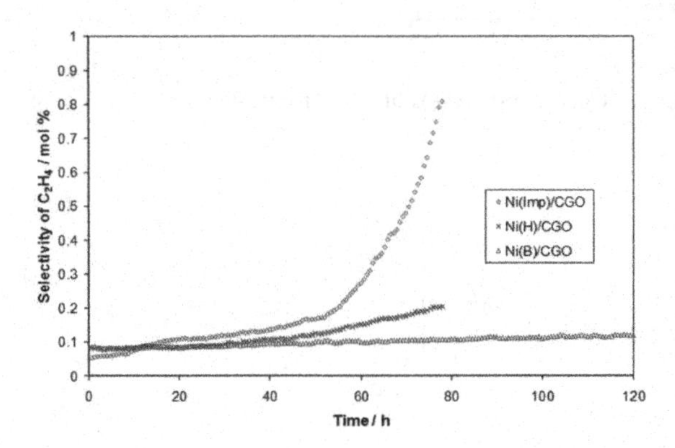

Figure 5. Selectivity of C_2H_4 in the ATR of the surrogate diesel catalyzed by different catalysts

Effect of Ni loading

As listed in table 1, the Ni(B)/CGO catalyst containing 1.5 wt % Ni was tested in ATR of dodecane. This catalyst showed very good conversion (>97%), high H_2 selectivity (>62%), and chemical resistance to carbon deposition. In the reforming of long chain alkanes, ceria has a high thermal cracking catalytic activity. It thus contributed to the conversion of decane to light hydrocarbon fragments, which subsequently underwent further reforming at the Ni sites to produce reformate.

Table 1. Conversion and product selectivity of reforming surrogate diesel with variation of its components over the Ni(B)/CGO catalyst*

Catalyst	Ni(B)/CGO 1.5 wt. % Ni	Ni(B)/CGO 1.5 wt. % Ni	Ni(B)/CGO 7.5 wt. % Ni	Ni(B)/CGO 7.5 wt. % Ni	Ni(B)/CGO 7.5 wt. % Ni	Ni(B)/CGO 15 wt. % Ni
Fuel	Dodecane	75 wt. % Dodecane, 25 wt. % Tetralin	75 wt. % Dodecane, 25 wt. % Tetralin	75 wt. % Dodecane, ~25 wt. % Tetralin, 20 ppmw MBTP	75 wt. % Dodecane, ~25 wt. % Tetralin, 50 ppmw MBTP	75 wt. % Dodecane, ~25 wt. % Tetralin, 50 ppmw MBTP
Conversion (%)	97.9	90.2	96.3	95.9	93.9	96.3
Product selectivity						
H_2 (mol %)	62.1	60.5	61.6	61.9	61.2	61.4
CO_2 (mol %)	22	25.3	23.3	23.5	23.8	23.6
CO (mol %)	10.6	9.95	11	10.8	10.7	10.9
CH_4 (mol %)	4.64	3.68	3.54	3.47	3.8	3.67

*The other reactor conditions remained unchanged.

When the liquid fuel contained aromatic compound, tetralin, in addition to dodecane, a drop in catalytic activity was seen because aromatic hydrocarbons are chemically more stable than aliphatic ones and the coking degree increases. Hence, inclusion of aromatic component in the proxy fuel could test tolerance of catalyst to the deposition of carbon. Clearly 1.5 wt % Ni content could not sustain the feed stream containing 25 wt% tetralin and with the same space velocity. Raising the Ni content to 7.5 wt. % could bring the conversion back to >96%, implying that a higher concentration of Ni catalytic sites over CGO was critical to deal with slow thermal cracking rates of aromatic compounds. On the basis of this, introduction of organosulfur compound, such as substituted thiophenes, into the surrogate fuel was an important testing criterion because commercial jet fuels contain high concentrations of organic sulfur compounds. They are the strong poisoning reagent to transition metal catalysts. In this test, MBTP was selected to conduct the S durability test. The same catalyst of Ni(B)/CGO containing 7.5 wt. % Ni was then employed to test the surrogate diesel containing different amount of MBTP. With the increase in S content, the conversion decreased. To improve the resistance to MBTP of 50 ppm in the fuel, the catalyst containing 15 wt % Ni was used, which improved conversion. Lastly, the ATR conversions of the surrogate diesel fuel comprising dodecane (75 wt. %), tetralin (~25 wt. %) and MBTP (50 ppmw) over the Ni(B)/CGO catalysts with different Ni contents was plotted in Fig. 6. The lowest nickel content of 1.5 wt. % showed a conversion of ~80%. The data were collated after 25 h reaction when the conversions became stable. It can be found from analyzing these assessment results that the deteriorating effects of aromatic and organosulfur compounds on the Ni catalyst are obvious. With the increase of Ni content, the conversion was improved until a stable level was achieved. We also tested the ATR of Esso Diesel fuel using the Ni(B)/CGO catalyst, a conversion of 85-90% and stable H_2 selectivity of 65% could be achieved.

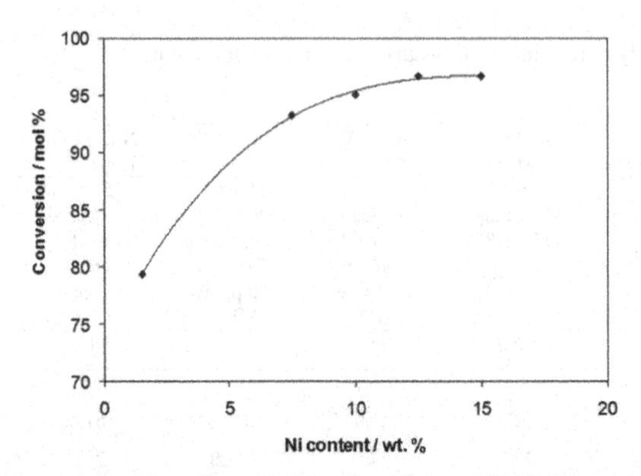

Figure 6. The effect of Ni loading in the Ni(B)/CGO on the conversion of the surrogate diesel

CONCLUSIONS

Compared with the impregnation catalyst, the catalysts prepared by the nanoparticle transfer method developed in this study, especially using KBH_4 as the reducing agent to produce nanosized Ni_{17}-B alloy particles initially, showed excellent catalytic activity, high resistance to carbon deposition and sulfur poisoning in the autothermal reforming of surrogate diesel comprising alphatic, aromatic and organosulfur compounds. This outcome can be attributed to a high dispersion of Ni particles in the CGO catalyst support and a special interface between Ni and CGO mediated by the B_2O_3 phase. The CGO catalyst support itself has thermal cracking activity to degrade heavy alkanes like dodecane at high temperatures. Besides this, it has obvious resistance to carbon deposition and sulfur poisoning as well. Finally, the initial boron component was converted to B_2O_3, whose role is presumed to be enhancing strong interactions between Ni and CGO.

ACKNOWLEDGMENT

This work was funded by the Defense Science and Technology Agency of Singapore.

REFERENCES

1. X. Zhang, S.H. Chan, G. Li, H.K. Ho, J. Li, Z. Feng, *J. Power Sources* **195** 685 (2010).
2. D. Shekhawat, D.A. Berry, T.H. Gardner, J.J. Spivey, *Catalysis* **19** 184 (2006).
3. M. Krumpelt, T. Krause, J.D. Carter, J.P. Kopasz, S. Ahmed, *Catal. Today* **77** 3 (2002).
4. B.D. Gould, X. Chen, J.W. Schwank, *J. Catal.* **250** 209 (2007).
5. B.D. Gould, X. Chen, J.W. Schwank, *Appl. Catal. A* **334** 277 (2008).
6. X. Chen, B.D. Gould, J.W. Schwank, *Appl. Catal. A* **356** 137 (2009).
7. M.C. Alvarez-Galvan, R.M. Navarro, F. Rosa, Y. Briceño, F.G. Alvarez, J.L.G. Fierro, Int. *J. Hydrogen Energy* **33** 652 (2008) .
8. M. Kobayashi, M. Flytzani-Stephanopoulos, Ind. Eng. Chem. Res. **41** 3115 (2002).
9. A.M. Azad, M.J. Duran, A.K. McCoy, M.A. Abraham, *Appl. Catal. A* **332** 225 (2007).

Mater. Res. Soc. Symp. Proc. Vol. 1446 © 2012 Materials Research Society
DOI: 10.1557/opl.2012.917

Shape-controlled Synthesis of Silver and Palladium Nanocrystals using β-Cyclodextrin

Gilles Berhault[1], Hafedh Kochkar[2] and Abdelhamid Ghorbel[2]
[1] Institut de Recherches sur la Catalyse et l'Environnement (IRCELYON), CNRS – University Lyon I, Villeurbanne, France
[2] Laboratoire de Chimie des Matériaux et Catalyse, University El Manar, Tunis, Tunisia

ABSTRACT

Shape-controlled Ag and Pd nanocrystals were synthesized using seed-mediated techniques using β-cyclodextrin (β-CD) as structuring agent. First, seeds were obtained by reacting Na_2PdCl_4 or $AgNO_3$ with a strong reducing agent, $NaBH_4$ in the presence of sodium citrate dihydrate acting as stabilizing agent. These seeds were then injected into a growth solution containing the same metallic precursor, ascorbic acid (as a weak reducing agent) and β-CD as structuring agent. TEM results emphasize a strong influence of the relative concentration of β-CD on the final morphology. In the case of silver, well-facetted nanocrystals were obtained with a progressive shift from kinetically to thermodynamically-controlled nanoobjects when increasing the β-CD/Ag molar ratio. On palladium, β-CD leads to the formation of dendrites (urchin-like) or multipods through controlled aggregation of primary particles. The use of β-CD does not interfere negatively with the catalytic properties of the Pd nanocrystals in the hydrogenation of cinnamaldehyde.

INTRODUCTION

Shape-controlled synthesis of metallic nanocrystals has become an important topic in materials science in the last 10 years due to a great interest in the unusual optical properties of anisotropic Ag and Au nanocrystals (NCs). Indeed, plasmonic properties strongly depend on the morphology of Au and Ag nanoobjects [1] leading to applications as biological and chemical sensors or as photothermal therapy agents.

Controlling the shape of metallic NCs can also be interested for catalytic applications since well-defined Ag, Pt, Pd or Rh nanoobjects exhibit preferential exposition of certain crystallographic facets. This allows to establish structure-reactivity relationships for structure-sensitive reactions like the selective hydrogenation of dienes or of α,β-unsaturated aldehydes [2] contrary to classical isotropic objects whose morphology is not clearly known. One general strategy to obtain well-defined nanoobjects is based on the selective adsorption of polymers or surfactants on specific facets leading to preferential growth along other directions. However, even if considerable progress has been obtained in the last years in the fine control of the nanomorphology of metallic NCs, only a partial understanding of the growth mechanism leading to the selective formation of a particular shape has been achieved. In this respect, the classical crystallization theory cannot be solely considered. Recent results acquired by some of us have shown that oriented attachment of primary particles also occurs mainly at the early stage during the growth of Pd NCs formed from a seed mediated approach in the presence of cetyltrimethylammonium bromide (CTAB) [3].

Another important issue to be solved for improving the colloidal approach used for synthesizing such nanoobjects is also to develop "greener" approach using biocompatible agents. In this respect, the replacement of high concentrations of alkyltrimethylammonium halides acting as surfactants by harmless compounds is required. One option envisaged in the present study is to use stable non-toxic cyclodextrin molecules in replacement of CTAB. Cyclodextrins form a group of cyclic oligosaccharides composed of six (α-CD), seven (β-CD) or eight (γ-CD) glucopyranose units. These compounds form truncated cones with cavities of 7.8 Å for instance for β-CD. These compounds are also amphipathic with a hydrophobic primary face and a hydrophilic secondary face. Up to now, β-CD was only envisaged for synthesizing Au NCs in combination with surfactants. The objective of the present study is therefore to evaluate the potential interest of β-CD as structuring agent for synthesizing Ag and Pd NCs.

EXPERIMENT

Pd and Ag seeds were first prepared through the rapid reduction of a Pd or Ag precursor in the presence of a strong reducing agent, $NaBH_4$. A 10 mL aqueous solution containing 0.25 mM $AgNO_3$ (or Na_2PdCl_4) and 0.25 mM sodium citrate dihydrate was first prepared. Next, 25 μL of ice-cold 0.1 M $NaBH_4$ was quickly injected into the solution while stirring vigorously for 2 h. Growth solutions were prepared by adding various amounts of 0.025 M $AgNO_3$ or 0.005 M Na_2PdCl_4 to an aqueous solution containing 0.005 M β-CD in order to monitor the β-CD/Ag(Pd) molar ratio from 50 to 150 for Ag and from 20 to 150 for Pd.

For the catalytic evaluation in the hydrogenation (HYD) of cinnamaldehyde, the Pd NCs were first deposited onto TiO_2 P25 support ($S_{BET} = 50\ m^2/g$) using classical impregnation techniques. 80 mg of the as-obtained Pd/TiO_2 catalysts were then added to 150 mL of isopropanol and to 400 mg of cinnamaldehyde and were then placed in a batch reactor working at 10 bars H_2 and at a temperature of 323 K. The course of the reaction was followed by analyzing samples using gas chromatography.

RESULTS AND DISCUSSION

Formation of Shape-controlled Ag NCs

In the case of silver, the use of β-CD leads to the formation of well-facetted NCs. Different morphologies were obtained whose proportion varied with the β-CD/Ag molar ratio. At a β-CD/Ag molar ratio of 50, a mixture of isosahedra, rods, triangular nanoplates, bipyramids or disk-like particles were obtained. However, when increasing the β-CD/Ag molar ratio to 150 leads to the selective formation of multiply-twinned icosahedral Pd NCs exposing mainly {111} facets (figure 1).

Morphology of NCs is known to depend on several experimental parameters (concentration and nature of stabilizing and reducing agents, amount and nature of seeds, concentration of metallic precursor) which can shift the growth regime from a kinetic to a thermodynamic control. While anisotropic NCs with higher surface areas are formed under kinetic control, equilibrium NCs with a "rounded" aspect and low energy facets are obtained under thermodynamic control. In this respect, increasing the rate of addition of monomers to the growing seeds leads to a shift from thermodynamic to kinetic nanoobjects. In the present case, increasing the β-CD/Ag molar ratio leads to a progressive shift to a thermodynamic growth regime. This shift can also be understood considering the fact that the final morphology of a nanocrystal depends on the stabilized structure of the growing seed. During the NC growth, twin defects are generally formed on the seeds leading to an extra strain energy compensated by a higher proportion of low surface energy {111} facets. If the growth rate becomes too fast, the extra strain energy due to twin defects cannot be compensated anymore leading to their evolution into single twinned or single crystalline seeds. In the present case, the growth rate was slowed down by the increasing relative concentration of β-CD leading to the formation of multiply-twinned seeds evolving into icosahedra.

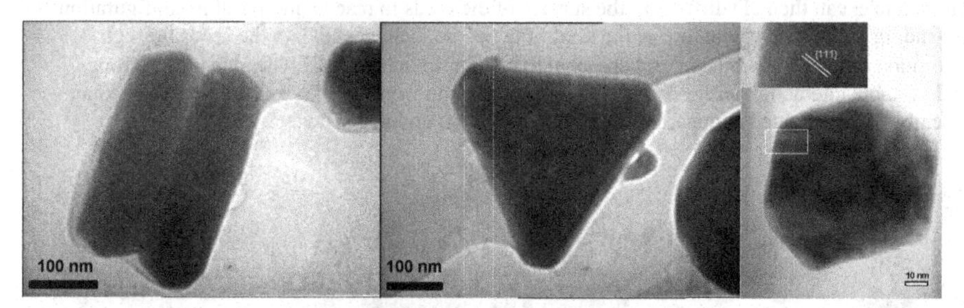

Figure 1. TEM images of Ag NCs obtained at a β-CD/Ag molar ratio of 50 (left: rods, middle: triangular nanoplate) and at a β-CD/Ag molar ratio of 150 (right: icosahedral, inset: HRTEM image of the white squared zone showing {111} fringes).

Formation of Shape-controlled Pd NCs

In the case of Pd, contrary to Ag, the interaction of β-CD does not lead to well-facetted objects but to the formation of dendrites (urchin-like) or multipods depending on the β-CD/Pd molar ratio (figure 2). The use of a β-CD/Pd molar ratio of 20 selectively leads to the formation of dendrites (urchin-like) formed of aggregated primary particles. HRTEM image of one of the branches of these dendrites (inset figure 2A) allows to observe distinct fringes corresponding to a d spacing of 2.3 Å close to the expected d_{111} value of 2.25 Å for fcc Pd. Increasing the β-CD/Pd molar ratio to 50 still leads to dendrites but with a more irregular morphology whereas some isolated NPs can also be visualized. These isolated NPs probably result from a too high stabilization by β-CD hindering their aggregation to dendrites. Further increase of the β-CD/Pd molar ratio to 100 leads to multipods formed mainly of 4 to 5 branches. Finally, the use of a β-CD/Pd molar ratio of 150 leads to ill-defined multipods and to isolated nanoparticles. Therefore, results show that β-CD regulates the final morphology of Pd NCs by restraining their final agglomeration leading successively to isolated nanoparticles, multipods, and finally dendrites with decreasing relative concentration of β-CD. It should be underlined that complementary experiments performed without using β-CD leads to isolated spherical nanoparticles. The restraining effect of β-CD on the growth of Pd NCs was also confirmed through the determination of the statistical size distribution of NCs in function of the β-CD/Pd molar ratio. Average size decreases from 47.7 ± 12.1 nm at a β-CD/Pd molar ratio of 20 to 35.4 ± 7.3 nm (β-CD/Pd ratio = 50) and finally to 20.9 ± 2.8 nm (β-CD/Pd molar ratio of 100).

The different morphologies achieved with β-CD for Pd compared to Ag can be partly ascribed to a higher hydrophobic-hydrophobic interaction of the narrow primary face of β-CD with Pd. It can also be explained considering an oriented attachment mechanism. Indeed, previous experiments performed using similar experimental conditions but using CTAB instead of β-CD have shown that Pd NCs are generally formed at the early growth stage through an oriented attachment process [3]. At the early growth stage, in the presence of CTAB, following a second heterogeneous nucleation induced by the injection of seeds into the growth solution, as-formed secondary nuclei aggregate to existing seeds by sharing their CTAB bilayer (figure 3). These nuclei can then 2D diffuse on the surface of the seeds to reach a more stable configuration depending on the size and nature of the seeds. Nuclei are then locked onto the seeds through resorption of the CTAB bilayer and epitaxial matching. In the case of β-CD, due to the strong hydrophobic interaction with Pd, the 2D surface diffusion would be blocked leading more to an assembly of aggregated nanoparticles than to an ordered oriented growth.

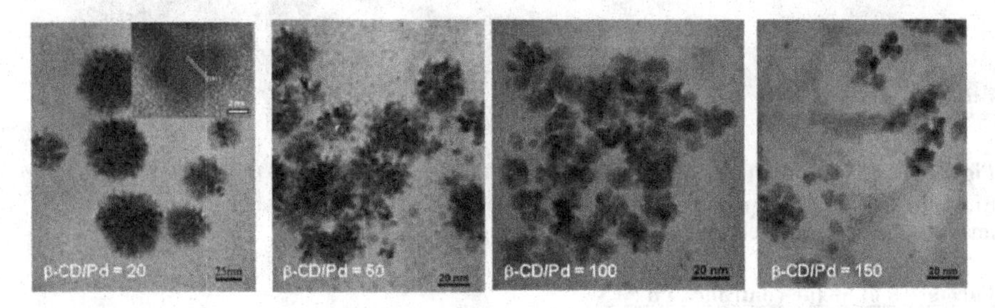

Figure 2. TEM images of the different Pd NCs (dendrites or multipods) obtained at β-CD/Pd molar ratios of 20, 50, 100, and 150. Inset left figure: HRTEM image of one of the branches of a dendrite showing the presence of {111} fringes.

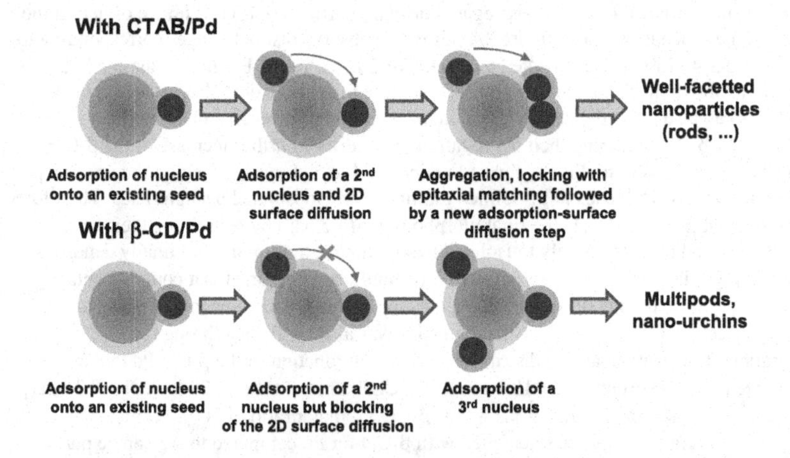

Figure 3. Representation of the oriented attachment mechanism occurring on growing Pd seeds in the presence of CTAB (top) or β-CD (bottom).

Cinnamaldehyde Hydrogenation over TiO_2-supported Pd NCs

As-formed Pd dendrites or multipods were deposited onto TiO_2 support and then evaluated in the HYD of cinnamaldehyde. This model structure-sensitive reaction of hydrogenation of a α,β-unsaturated aldehyde was used to evaluate the preferential hydrogenation of either C=O or C=C bonds on catalytic systems and it provides through variation of selectivity properties valuable information about the type of facets exposed to the reactant. On Pd catalytic systems, the hydrogenation of C=C bonds is favored leading to a reaction scheme formed of two parallel pathways, one corresponding first to the hydrogenation of the C=C bond leading to the formation of hydrocinnamaldehyde (HCALD) and another pathway leading first to the hydrogenation of the C=O bond and then formation of cinnamylic alcohol (CA) followed by rapid hydrogenation of the remaining C=C bond into 3-phenyl-1-propanol (PP). Pd/TiO_2 catalysts were then synthesized by using Pd NCs formed at three different β-CD/Pd molar ratios and were labeled Pd(x)/TiO_2 with x = β-CD/Pd molar ratio. Pd metal loading was fixed at 0.2 wt%. Initial rate constants were found identical for Pd(20)/TiO_2 and Pd(50)/TiO_2 while a lower value was obtained for the Pd(100)/TiO_2 catalyst (table 1). However, these values only reflect the size of the primary particles constituting the dendrites or multipods. While similar values were found for Pd(20)/TiO_2 and Pd(50)/TiO_2 (respectively 6.2 ± 1.3 nm and 6.9 ± 1.0 nm), bigger primary particles (10.1 ± 1.2 nm) forming multipods were obtained on Pd(100)/TiO_2 leading to a lower surface-to-volume ratio and therefore a lower density of active sites per gram of catalyst. Turnover frequency values (TOFs) were also determined for each catalyst in order to evaluate the intrinsic activity per Pd site. Results show relatively similar values (1.7-2.1 s^{-1}) whatever the β-CD/Pd molar ratio used to synthesize the Pd NCs. These values were also similar to a Pd/TiO_2 reference prepared without using β-CD (1.4 s^{-1}). These results show that the presence of β-CD at the surface of the Pd particles does not interfere negatively with the catalytic properties of the final nanostructured catalysts. This last point is quite important since to be applied for catalytic applications, shape-controlled metallic nanocrystals must be obtained using structure-directing agents which can be removed easily before catalytic application or which cannot poison active sites irreversibly. The present results validate the use of β-CD as a harmless structure-directing agent in a catalytic point of view. This oligosaccharide molecule can then be envisaged for preparing metallic NCs presenting a potential interest in catalysis.

Table 1. Pd loadings, initial rate constants, turnover frequency values and selectivity results (at 20% conversion) for the different Pd(x)/TiO_2 catalysts.

Catalyst	Pd Loading (wt%)	Initial Rate Constant (10^{-6} mol.s^{-1}.g^{-1})	TOF (s^{-1})	Selectivity HCALD	Selectivity PP	Selectivity CA
Pd(20)/TiO_2	**0.21**	**6.0**	**1.7**	49	50	1
Pd(50)/TiO_2	**0.19**	**6.1**	**2.1**	51	47	2
Pd(100)/TiO_2	**0.20**	**3.8**	**1.8**	46	51	3

Pd(x)/TiO_2 catalysts with x = β-CD/Pd molar ratio used initially

About selectivity, results also appear relatively similar whatever the Pd/TiO_2 catalyst. Selectivity values close to 50 % were obtained for hydrocinnamaldehyde and 3-phenyl-1-

propanol. These values are close to those expected from theoretical calculations for Pd(111) confirming TEM results about the main exposition of {111} facets on these Pd NCs [4].

CONCLUSIONS

β-cyclodextrin was used successfully to synthesize shape-controlled Ag and Pd NCs using a seed-mediated approach. Depending on the nature of the metallic NCs to be formed, different kinds of nanostructures were obtained. In the case of silver, using β-CD led to the formation of well-facetted nanoobjects, mainly icosahedra while in the case of palladium, dendrites or multipods were formed by the assembly of primary particles. For Ag, the role of β-CD is to regulate the size of the growing seeds favoring the stabilization of twinned seeds evolving into icosahedra. With Pd, the stabilization effect induced by β-CD restrains the synthesis to morphologies resulting from the attachment of primary particles. This study finally emphasizes that β-CD can be used to obtain shape-controlled supported metallic catalysts intrinsically active in hydrogenation reactions without interference with the catalytic properties of these metallic nanocrystals.

ACKNOWLEDGMENTS

This work has been supported by the "Action Intégrée Franco-Tunisienne du Ministère des Affaires Etrangères et Européennes Français et du Ministère de l'Enseignement Supérieur de la Recherche Scientifique et de la Technologie Tunisien".

REFERENCES

1. J. Perez-Juste, I. Pastoriza-Santos, L. M. Liz-Marzán and P. Mulvaney, *Coord. Chem. Rev.* **249**, 1870-1901 (2005).
2. L. Piccolo, A. Valcarcel, M. Bausach, C. Thomazeau, D. Uzio and G. Berhault, *Phys. Chem. Chem. Phys.* **10**, 5504-5506 (2008).
3. L. Bisson, C. Boissière, L. Nicole, D. Grosso, J. P. Jolivet, C. Thomazeau, D. Uzio, G. Berhault and C. Sanchez, *Chem. Mater.* **21**, 2668-2678 (2009).
4. F. Delbecq and P. Sautet, *J. Catal.* **152**, 217-236 (1995).

AUTHOR INDEX

Abdi, Fatwa F., 7
Alves, Annelise K., 53
An, Woo-Jin, 71

Bergmann, Carlos P., 53
Berhault, Gilles, 89
Berutti, Felipe A., 53
Biswas, Pratim, 71
Braun, Artur, 19
Brooks, Christopher, 65
Bugayong, Joel, 59

Caballero, Alfonso, 47
Cabrera, Carlos R., 39
Chang, Yuancheng, 19

Dabirian, Ali, 7
DeAngelis, Alexander, 19
Deura, M., 1
Dunn, Steve, 13

Firet, Nienke, 7
Fong, Hao, 77

Gaillard, Nicolas, 19
Gangopadhyay, Shubhra, 71
Ghorbel, Abdelhamid, 89
Gonzalez-Delacruz, Victor M., 47
Griffin, Gregory L., 59

Hayashi, T., 1
Hoa, Le Q., 33
Holgado, Juan P., 47
Hong, Liang, 83
Hu, Boxun, 65

Karabacak, Tansel, 25
Kariuki, Nancy, 25
Kerce, Kimberly, 77
Khudhayer, Wisam J., 25
Kochkar, Hafedh, 89
Kolla, Praveen, 77
Kreidler, Eric, 65

Liu, Zhaolin, 83

Mayol, Ana-Rita, 39
Menéndez, Christian L., 39
Mukherjee, Somik, 71
Myers, Deborah J., 25

Niedzwiedzki, Dariusz M., 71

Ohara, W., 1
Ohkawa, K., 1

Pereñiguez, Rosa, 47

Ramalingam, Balavinayagam, 71
Reolon, Raquel P., 53

Saito, Masato, 33
Shaikh, Ali U., 25
Smirnova, Alevtina, 77
Stock, Matt, 13
Suib, Steven L., 65

Tamiya, Eiichi, 33
Ternero, Fatima, 47
Tsai, Cheng-Chun, 59

Uchida, D., 1

van de Krol, Roel, 7

Wang, Wei, 83
Wang, Wei-Ning, 71

Xia, Zetao, 83

Yoshikawa, Hiroyuki, 33

SUBJECT INDEX

aerogel, 77
Ag, 89

catalytic, 13, 33, 39, 47, 53, 65, 77, 83
Co, 65
crystallographic structure, 19
Cu, 59

electrical properties, 19
electrochemical synthesis, 59
electrodeposition, 39
energy generation, 7, 33
energy storage, 59, 83
extended x-ray absorption fine
structure (EXAFS), 47

ferroelectric, 13
fiber, 53

H, 1

III-V, 1

methane, 53
Mn, 65

nanostructure, 39, 71, 77, 89
Ni, 83

Pd, 89
photochemical, 1, 7, 13, 19, 71
polymer, 33

spray pyrolysis, 7

thin film, 71

x-ray photoelectron spectroscopy
(XPS), 47

Printed in the United States
by Baker & Taylor Publisher Services